The Natural World and Science Education in the United States

Ajay Sharma • Cory Buxton

The Natural World and Science Education in the United States

palgrave
macmillan

Ajay Sharma
Department of Educational
Theory and Practice
University of Georgia
Athens, GA, USA

Cory Buxton
University of Georgia
Athens, GA, USA

ISBN 978-3-030-09424-9   ISBN 978-3-319-76186-2 (eBook)
https://doi.org/10.1007/978-3-319-76186-2

© The Editor(s) (if applicable) and The Author(s) 2018
Softcover re-print of the Hardcover 1st edition 2018
This work is subject to copyright. All rights are solely and exclusively licensed by the Publisher, whether the whole or part of the material is concerned, specifically the rights of translation, reprinting, reuse of illustrations, recitation, broadcasting, reproduction on microfilms or in any other physical way, and transmission or information storage and retrieval, electronic adaptation, computer software, or by similar or dissimilar methodology now known or hereafter developed.
The use of general descriptive names, registered names, trademarks, service marks, etc. in this publication does not imply, even in the absence of a specific statement, that such names are exempt from the relevant protective laws and regulations and therefore free for general use.
The publisher, the authors and the editors are safe to assume that the advice and information in this book are believed to be true and accurate at the date of publication. Neither the publisher nor the authors or the editors give a warranty, express or implied, with respect to the material contained herein or for any errors or omissions that may have been made. The publisher remains neutral with regard to jurisdictional claims in published maps and institutional affiliations.

Cover design by Ran Shauli

Printed on acid-free paper

This Palgrave Macmillan imprint is published by Springer Nature
The registered company is Springer International Publishing AG
The registered company address is: Gewerbestrasse 11, 6330 Cham, Switzerland

ACKNOWLEDGMENTS

We would like to acknowledge the many people who have helped us during the creation of this book. First and foremost are our spouses, Venu and Jean-Marie, who are always supportive of our passion for this work and the long hours that it requires. Our children, Dhruv, Tanvi, Jonah, Remy, and Lindy, remind us on a daily basis why we are committed to improving science education in ways that will help future generations to better understand the relationships between humans and the rest of the natural world.

Special thanks go to our colleagues in the Department of Educational Theory and Practice at the University of Georgia, who continually push us to think in new ways. A big shout-out especially to our dear friend and colleague Mardi Schmeichel for her continuous encouragement and prodding that was critical during the writing of the chapters. The initial research for the book was made possible by a faculty research grant from the University of Georgia. We are also grateful to the University of Georgia's College of Education for giving one of us (Sharma) a semester-long study leave that enabled us to write this book with minimal distractions or hardships.

A work of this nature inevitably builds on ideas and research of its predecessors. The genesis and inspiration of research reported in this book can be directly traced to the Environmental Literacy Project led by Charles W. (Andy) Anderson at the Michigan State University. We also thank Andy for giving much valuable feedback on one of the chapters in this book. Thanks are also due to Jeffrey Hepinstall-Cymerman at the University of Georgia for being our go-to expert on the latest developments in ecology. He was always ready to help and his expertise was invaluable in the development of the conceptual framework we have proposed for science education in this book. We are also grateful to our colleague Ruth Harman for her generous and valuable comments, criticism, and advice on how to correctly use systemic functional linguistics to do critical discourse analysis of texts.

Special thanks are also owed to the teacher, students, and school where much of the empirical data used in this book were collected. While Ms. Gilmour and Little Creek Middle School are pseudonyms, the actual teacher, students, and school were gracious and thoughtful participants, and this book would not be possible without their willing participation.

Milana Vernikova, our editorial assistant at Palgrave Macmillan, was patient and insightful, and her help throughout made this book a better product. Our sincere thanks also go to the Production Editor (Arun Prasath) and the copyeditors (Priti Debolina and R. De Guzman) for their help in finalizing the many details that go into completing a project of this kind. Finally, we would like to thank the reviewers who gave us thoughtful feedback and suggestions during the proposal process.

We learned a great deal during the writing of this book and hope that it will be of some value to you, the reader. We can then work together to better prepare our students to survive and thrive with justice and equity in the Anthropocene Epoch.

Athens, GA, USA, 2017   Ajay Sharma
                                                                                      Cory Buxton

Contents

1 Introduction 1

2 Evolving Views on the Nature of Nature 21

3 The Intended Curriculum: Locating Nature in the Science Standards 45

4 The Intended Curriculum: Nature as Represented in a Science Textbook 87

5 The Enacted Curriculum: Representations of Nature in Science Teaching 121

6 The Received Curriculum: Nature as Understood by Students 149

7 A Sustainability Science-Based Framework for Science Education 169

Index 207

List of Figures

Fig. 3.1 Next Generation Science Standards   54
Fig. 3.2 Georgia Performance Standards   56
Fig. 3.3 Georgia Standards of Excellence   58
Fig. 7.1 Sustainability science-based framework for science education   187

List of Tables

Table 4.1	Chapter outlines	98
Table 4.2	Clauses involving humans/social world	99
Table 4.3	Agency analysis in material process clauses	102
Table 4.4	Positive and negative references to major actors in environmental problems and solutions	106
Table 5.1	Unit sequence	130
Table 5.2	Warm-up questions	131
Table 5.3	Examination questions	132
Table 5.4	Environmental science unit plan essential questions	133
Table 5.5	Questions in the review game	136

CHAPTER 1

Introduction

In 2007, environmental writer Bill McKibben asked climate scientist James Hansen what the safe limit would be for carbon dioxide concentration in the atmosphere (Monastersky, 2009). Hansen took some time to do the calculations and came back with a target figure of 350 parts per million (ppm) that humanity should aim not to exceed if it "wishes to preserve a planet similar to that on which civilization developed and to which life on Earth is adapted" (Hansen et al., 2008; p. 217). That was then, a moment in human history when one could still be confident about the planet's future without sounding naïve and out-of-touch. 350 ppm of atmospheric carbon dioxide as a target figure now certainly looks like a wishful thinking for a future that would never be. In April 2017, atmospheric concentrations breached the 410 ppm threshold (Kahn, 2017). How dangerous the current trajectory of growth in carbon dioxide emissions is can be estimated by a recent report that predicts that there is now only a 5% chance that we will be able to meet the Paris Climate Accord's aspirational goal to keep the global temperature rise below the widely perceived critical tipping point of 2 degrees Celsius (Raftery, Zimmer, Frierson, Startz, & Liu, 2017). As if the threat from runaway climate change was not enough, the planet is being continuously buffeted by a growing number of other grave ecological crises, such as the ongoing sixth mass extinction and the increasing scarcity of fresh water, that have already begun to wreak havoc on the lives of marginalized human communities as well as numerous other species (Barnosky et al., 2011; Pearce, 2006). As a result, scientists have begun wondering if a global collapse of human civilizations is imminent (Ehrlich & Ehrlich, 2013). After all, ecological problems have been the prime culprits in the collapse of several civilizations in the past (Diamond, 2013). Many people in the United States and other industrialized nations may find such talk unduly alarmist because when they look around they see few obvious signs of ecological distress. But civilizational collapse doesn't happen in a year or even a decade. It took more than 400 years for the Mayan civilization to collapse and many more centuries for the Indus Valley civilization. We would do well to remember that industrialized societies have been around for only about 300 years, and the alarm bells being rung by the scientific community for imminent danger to human existence are already too loud to ignore.

The good news, as we explore this crisis, is that it is not just the scientific community that is worried about the fate of the planet. Worrying about the ecological health of the planet has also become a constituent part of the precarity of life for all of us living in the current Anthropocene Epoch. For instance, in a recent opinion poll, about three-quarters of US adults were concerned about the environment and wished that the country did whatever it took to protect it (Anderson, 2017). A lay environmentalism has indeed become a core value that most people in the United States can be said to share (Dietz, Fitzgerald, & Showm, 2005; Kempton, Boster, & Hartley, 1995; Sellers, 2012). Though, as Turin (2014) suggests, "environmental concerns, which on a spectrum of political goods might be considered 'third-order' goods, are particularly susceptible to being overwhelmed by other issues possessing more immediate and tangible impacts" ("Conclusion", para. 47). However, it cannot be denied that, when asked, US adults convey a broad support for taking environmental action, both at the individual and institutional level, to overcome ecological challenges.

In times of economic security, general peace, and prosperity, when US citizens are not feeling economic distress and don't have pressing concerns about their security, this support has the potential to lead to effective systemic changes that tackle ecological challenges. But what if people's environmental concerns and support for action rest on mistaken assumptions and inadequate understandings about the world? Does that make us inclined to support ideas and actions that are at best benignly ineffective? These questions are pertinent because research indicates a broad support for environmental actions that have been shown to be of dubious value in terms of their effectiveness. For instance, a clear majority of US population believes that new technology will solve current and future ecological problems, such as climate change. As a result, there is a broad support for technical and technocratic solutions in the United States (Funk & Kennedy, 2016; Turin, 2014). However, it has been clear for a long time that ecological issues, such as climate change, are, at their core, societal issues that have resulted from the current political situation, the dominant economic system, and the anthropogenic transformation of the biosphere since the industrial revolution (Brulle, 2000; Clark & York, 2005; Steffen, Crutzen, & McNeill, 2007; Turin, 2014). As a result, technical solutions on their own are highly unlikely to be effective in tackling our ecological woes. For instance, there is strong support at the policy level, in the marketplace, and among the public for planting trees to soak up the excess atmospheric carbon dioxide and thus mitigate the impact of climate change (Bäckstrand & Lövbrand, 2006; McGrath, 2017; Melnick, Pearl, & Warfield, 2015). However, scientists now generally agree that removal of carbon dioxide through afforestation is a failed strategy. For instance, a recent paper calculated the extent of terrestrial carbon dioxide removal through planting of trees that would be needed if we are not able to achieve reduction in emissions of carbon dioxide to prevent global mean temperature rise of 2.5% above pre-industrial level (Boysen et al., 2017). The scientists found that planting of trees or other ways of managed biomass growth will be "unable to counteract 'business-as-usual' emissions without eliminating virtually all natural ecosystems" (p. 463). In fact, the study concluded that even if considerable emissions reductions are assumed, terrestrial carbon dioxide removal strategies will require "> 1.1 Gha of the most productive agricultural areas or the elimination of >50% of natural forests" (p. 463). Similarly, there is a distinct worry among the scientists that geoengineering solutions might end up harming rather than helping our efforts to tackle climate change (Stephen, 2016; Shepherd, Iglesias-Rodriguez, & Yool, 2007).

Further, many people in the United States have come to see environmentalism as an individual personal virtue rather than as a matter of collective public action (Crowell & Schunn, 2014; Dryzek, 2013; Treanor, 2010). As a result, the approximately 20% US adults who actually do take regular actions to protect the environment, rarely go beyond individual actions, such as using re-usable shopping bags for trips to the grocery store (Funk & Kennedy, 2016). Here we clearly see a mismatching of ecological and sociological scales between the problems and the solutions practiced by individuals. This is because while personal actions, such as recycling, are local in scale, the ecological problems that bedevil our planet occur on local as well as larger regional, national, and international scales—both socially and ecologically. Research shows that individual environmental actions can be effective only when they are situated in the context of just and democratic governance of ecological resources (Dietz, Ostrom, & Stern, 2003; Hempel, 1996). In the absence of fair and democratic management of ecological commons, individuals have more to lose and little to gain through reduce, reuse, and recycle types of virtuous environmental actions—a situation that invariably leads to unsustainable exploitation of common property resources resulting in what has been called the *tragedy of commons* (Hardin, 2009).

People's support for ineffective and misdirected environmental solutions likely originates from a complex and situationally contingent interplay of diverse social, cultural, political, and economic factors that operate on multiple scales and material-discursive planes. One of the key factors in this complex causal web is people's (mis)understanding of the nonhuman world and their relationship with it. Environmental sociologists argue that there exists a dominant worldview, referred to as the *Dominant Social Paradigm (DSP)*, in Western industrialized societies, such as the United States, that reflects and shapes people's understanding on human's place in the world. This worldview is defined by the following themes:

- Low evaluation of nature for its own sake.
- Compassion mainly for those near and dear.
- The assumption that maximizing wealth is important and risks are acceptable in doing so.
- The assumption of no physical ("real") limits to growth that can't be overcome by technological inventiveness.
- The assumption that modern society, culture, and politics are basically okay (Harper & Snowden, 2017).

This *Dominant Social Paradigm* derives much of its support and legitimacy from the currently dominant discourses in the West, such as neoliberalism, scientific and conservative Judeo-Christian religious discourses that naturalize human exceptionalism, and the commodification of nature. This paradigm also corresponds closely with people's rather limited or incorrect scientific understanding of the basic ecological processes and phenomena that govern life on Earth (Ranney & Clark, 2016; Jenkins,

2003). In fact, a study showed that even highly educated adults, such as the graduate students at the Massachusetts Institute of Technology, failed to conserve matter and showed other widespread misunderstanding in their explanations of the fundamental processes regarding greenhouse gas emissions (Sterman & Sweeney, 2007). Unsurprisingly, currently only about 28% of US adults can be said to be civic-scientifically literate in terms of their ability to "find, make sense of, and use information about science or technology to engage in a public discussion of policy choices involving science or technology" (Miller, 2016; p. 2). This rate of scientific illiteracy may not look bad in terms of international comparisons (National Science Board, 2016). But, we need to do far better if we wish to see the needed degree of public support and activism for democratic and socioecologically just ways to face ecological challenges in the Anthropocene.

Of course, people's understanding of the world and their ways of talking about it evolve from diverse origins. As participants in multiple local, global, and "glocal" material-discursive networks, we all actively as well as passively imbibe knowledge, ideas, attitudes, folklore, and perspectives about the environment and our relationship with it from a vast array of sources (Eagles & Demare, 1999; Weaver 2002). However, it cannot be denied that school science, as a carrier of official, authoritative knowledge, constitutes one of the critical formative influences that shapes how and what we come to understand about our world (Tikka, Kuitunen, & Tynys, 2000; Weaver, 2002). Thus, if we wish to see greater public support for ideas, practices, and policies that are scientifically accurate and socioecologically just, one important facet to explore is the representations of the world that are conveyed, legitimized, and naturalized by school science. We are concerned that science as currently prescribed, taught, and learned in US schools may not be preparing students well to understand and do something meaningful about the complex and calamitous ecological crises facing our planet, such as climate change and ongoing mass species extinction (Assaraf & Damri, 2009; Covitt, Tan, Tsurusaki, & Anderson, 2009). Thus, we undertook a multidimensional investigation to explore the contours of the intended, enacted, and received curricula in the United States on ecology, environmental, and Earth science topics that are central to our understanding of the natural world and our relationship with it. This book presents the key results of our investigation.

Despite a rich history of research on how different aspects of nature are taught and learned in classrooms, insufficient attention has been paid to how the natural world as a whole is represented and understood in science education. The last notable work addressing this topic was done about 15 years ago by William Cobern (2000). Adopting a worldview theoretical perspective, Cobern investigated ninth-grade students' and their science teachers' understandings about nature and the natural world. The focus of this work was on exploring the fundamental beliefs that students and teachers held about nature. Our book is different from Cobern's work in three important ways: (a) First, it is more current than Cobern's book, updating the topic to include the remarkable changes that have occurred in this field since the start of the new Millennium. (b) Second, our book has a more extensive focus, as it covers science curricula and instruction, as well as students' understanding of nature. (c) Finally, this book not only critiques but also presents an alternative sustainability science-based conceptual framework for science education that is compatible with the current understanding of our world in ecological and environmental sciences, explicit in its ethical stance, and oriented towards praxis in service of social-ecological justice.

A clarification is in order at this point. Following the current understanding among ecologists, we believe that our planet cannot be adequately and scientifically understood by dividing it into two distinct domains of natural world and social world, or nature and society. Rather, the world should be seen as an overlapping patchwork of materially open, nested, and coupled socioecological systems that are just as social as they are natural (Schmitz, 2016). However, as we report in this book, much of the science education community continues to refer to nature/natural world/natural systems to situate the normative terms, concepts, and explanations related to ecology and environmental science topics in intended, enacted, and received science curricula. The use of the etic term social-ecological system in place of the emic words nature/natural world/natural systems is likely to create a misunderstanding about the intended meaning for readers. Hence, in our exploration of school science we have preferred to use words like nature/natural world/natural systems in the same way as is currently prevalent in science education. We urge our readers to see our choice as an analytic device needed to better understand science curricula in its own terms and not as our implicit acknowledgment of its scientific validity.

Natural World in the School Science

As we describe in subsequent chapters, there is currently a burgeoning interest among science educators to understand how "Nature" (as in the natural world or environment) is taught and learned in science classrooms. But on the whole this still remains a little-explored area of research in science education. It is possible that this lack of research could be partly due to the remarkable and substantial outsourcing of efforts to understand teaching and learning about nature to the specialized subfield of environmental education research. Indeed, by broadening our search to include environmental education and other educational research journals, we did find more of a research base. We believe this to be related to the fragmentation of the science curricula across different conceptual areas, each claimed and protected by a community of researchers who bound their focus areas through membership and publication in certain community-specific journals. For instance, as Gruenewald (2004) noted some years ago, and as our own literature review revealed as well, it looks like the research on socioecological issues has been staked out by the environmental education community. Of course, there still exists some overlap among science education and environmental education researchers, but it seems clear to us that this siphoning off of most research on socioecological topics to specialized environmental education journals does impoverish other areas of research in science education. This practice also reifies the fragmentary nature of science curricula in the United States by encouraging omission and marginalization of socioecological topics that naturally cut across disciplinary boundaries. As evidence of this, we note that most of the relevant examples we found of research programs that highlight socioecological topics in science curricula came from the United Kingdom, Australia, and a few other countries, with little of this work occurring in the United States. We are not sure why this is so, and feel that this gap needs to be investigated, explained, and rectified. To compensate for the lack of research situated in the United States, wherever needed, we have chosen to include research done in other developed societies with the understanding that international research can be useful in providing indications of what research might have to say if conducted in US schools.

Learning about the biophysical world and our relationships with it is an important component of school science. Starting in the elementary grades, students do typically begin to learn about the biophysical world through basic ecology concepts, such as habitat and food chain, and about our relationship with that world through environmental issues, such as pollution and habitat destruction. There is also a relatively robust scholarship on socially relevant and ecologically sensitive issues by proponents of Science/Technology/Society (STS), Socioscientific Issues (SSI), ecojustice, citizen science, and place-based science education. One might expect that this kind of research would be interested in exploring representations of nature in school science and how nature gets taught and understood in science classrooms. However, we find that this is rarely the focus of research in STS, SSI, or ecojustice-focused science education research.

Instead, this body of research has largely focused on highlighting the importance of scientific topics with social ramifications as a context for reaching some desired goals in science education, such as greater student interest in learning science and better understanding of science content, or on understanding the nature of science and associated science practices, such as scientific argumentation and inquiry (Pedretti & Nazir, 2011). We believe that this research does tackle important issues, such as the historical and sociocultural contexts of scientific ideas and scientific work, evidence-based and ethico-moral decision-making on socioscientific issues, critiquing or solving socioecological problems through human action, indigenous knowledge systems, and place-based learning, especially as they relate to science education (Mueller & Tippins, 2012; Pedretti & Nazir, 2011; Sadler & Dawson, 2012). However, for research focused on teaching and learning about nature, we had to look mostly into environmental education research journals.

As we read the existing research in both science and environmental education journals, we couldn't help but notice the overwhelming tendency of this work to take as a given the premise that science content taught in schools faithfully reflects the current state of the field in disciplines such as ecology and geography. This research may raise questions regarding the inclusion or exclusion of certain concepts, such as climate change, but by and large, fails to raise broader issues such as the need to shift our focus in school science from ecological systems to socioecological systems perspectives. For instance, a group of researchers interested in environmental/ sustainability education recently outlined four expected future directions of research in their field (Aguayo et al., 2016). None of these directions involved a critique or re-envisioning of the overall ecological perspective that determines both the form and content of instruction on ecological topics in science or environmental education.

There are some notable exceptions, of course, that have tried to expand and re-envision science content on ecological issues, such as the research done by the research group led by Charles W. Anderson at the Michigan State University (http://envlit.educ.msu.edu/).[1] Research and advocacy done by this group on learning progressions and student understanding on several key concepts related to socioecological systems has helped the science education community to appreciate the need to broaden the focus in school science from ecological systems to socioecological systems. The Michigan State University group has been active in this field of inquiry for more than a decade, but they are no longer plowing a lonely furrow. We are encouraged to see that in the last few years, a small group of researchers has emerged in the United States who are actively engaged in critiquing the normative science content and offering alternative proposals as they relate to ecological concepts and the biophysical world.

The nature of this emergent critique and research foci can perhaps be best illustrated by the following three research studies. First, Feinstein and Kirchgasler (2015) did a discourse analysis of the Next Generation Science Standards (NGSS) to examine "how the NGSS explicitly define and implicitly characterize sustainability" (p. 121). They found that the scientific discourse in these standards is characterized by universalism, scientism, and technocentrism that combine to give a distinct technology-centered perspective to the issue of sustainability as presented in the science standards. This study presents a forceful argument that "students who are taught to think about sustainability from this perspective will be less able to see its ethical and political dimensions and less prepared for the political realities of a pluralist, democratic society that must balance the needs of multiple groups and integrate science with other sources of knowledge to develop contextualized responses to sustainability challenges" (p. 121). Second, another discourse analytic study on NGSS looked at how environment and environmental issues are conceptualized and positioned in these standards (Hufnagel, Kelly, & Henderson, 2017). The study found that the NGSS conceptualizes environment as an entity that is separate from people. The authors critique these science standards for their focus on technoscientific solutions and misrepresentation of the agency of actors in environmental issues. For our third example we have Ruppert and Duncan's (2017) Delphi study on the concept of Ecosystem Services—the benefits that humans derive from environmental systems—and the importance of teaching this concept in science classrooms. This study is notable because, going beyond critique, it proposes that we should teach about ecosystems and ecosystem services to students with the help of "a refined model of coupled human-environment systems that articulates multiple human populations as embedded within ecosystems, connected to these ecosystems near and far, and benefiting from the resources and conditions provisioned by these ecosystems" (p. 737).

As these three examples show, much of the new energy and attention in research on how we teach about our biophysical world in our science classrooms appears to be focused on exploration and improvement of the intended curriculum. Interestingly, researchers outside the United States have been active far longer in critiquing the normative science content taught in classrooms (see, e.g., Colucci-Gray, Camino, Barbiero, and Gray (2006), Hovardas and Korfiatis (2011), Lefkaditou, Korfiatis, and Hovardas (2014), and Lindahl and Linder (2015)). Because this kind of research is yet to gain widespread attention, we found that most science and environmental education researchers in the United States still subscribe to either a tacit or an explicit belief in a social-nature dualist ontology for describing the world, assuming that a balance of nature metaphor is still a good way to understand how natural systems regulate themselves to maintain homeostasis. For instance, Cheng and Monroe (2012) developed a "connection to nature" index to measure children's affective attitude towards nature, and Liu and Lin (2014) assumed that balance of nature is still a valid scientific concept.

As for the foci of research on nature within the ambit of science and environmental education, most researchers appear to be interested in students' conceptions of nature and human-nature relationships. Leaving aside a few noteworthy exceptions, such as Cobern's (2000) investigations of ninth-grade students' conceptualizations of nature (described above) and the work done by the research group at the Michigan State University (see, for instance, Covitt et al. (2009) and Tsurusaki and Anderson (2010)), much of this research has been done outside the United States. Based on this research we can make the following tentative observations about students' understanding of nature or the natural world. First, students seem to hold multiple conceptions of nature that they use in a contingent, opportunistic manner to enact their subject positions as students (Cobern, 2000; Nielsen, 2012; Pointon, 2014). These diverse conceptions likely reflect different perspectives such as religious, scientific, aesthetic, and conservationist orientations, based on students' experiences with the nonhuman world that they acquire over the course of their lives both in and beyond school. Further, it appears that beyond school discourses are more influential in shaping students' conceptions than are their experiences with school science (Covitt et al., 2009). This may explain why research finds that the link between factual science content and students' use of nature for participation in classroom discourse is generally weak (Nielsen, 2012; Pointon, 2014). Second, despite their eclecticism in conceptions of nature, most students exclude humans when conceptualizing nature or the natural world (Li & Ernst, 2015). This may be true for adults as well (Vining, Merrick, & Price, 2008). That is, nature is believed to be out "there" in a pristine, pure state as an object of fantasy and desire though it is often soiled by human interference (Lindahl & Linder, 2015; Payne, Cutter-Mackenzie, Gough, Gough, & Whitehouse, 2014). Third, students tend to see the world as largely orderly and harmonious, resulting in a strong belief that the natural world is marked by a balance of nature (Ergazaki & Ampatzidis, 2012; Hansson, 2014). When probed further on this, however, it also looks like most students do not actually understand the idea of balance of nature (Liu & Lin, 2014; Zimmerman & Cuddington, 2007). Finally, American students appear to be moving towards the idea that humans are stewards rather than exploiters of the natural world (Li & Ernst, 2015). Of course, as is the case with other concepts in ecology, students' understanding of the connections between the human and natural world is usually weak and patchy (Tsurusaki & Anderson, 2010).

In addition to research on students' ideas about nature, researchers have also explored curricular and instructional issues related to ecology topics in science and environmental education. Spurred by the rising threat of climate change and research showing that most students graduate from schools with a very deficient understanding of climate change, one particular area that has seen increased activity in recent years has been in climate change education. Over the last decade, some researchers and science educators have devoted considerable efforts to research on teaching and learning about climate change and developing conceptual frameworks and pedagogical toolkits to make climate change an important topic in science education and to enhance climate change literacy of US students (see, for instance, Shepardson, Roychoudhary, & Hirsh, 2017). The overall ecology framework that one sees in this kind of work is of systems ecology with a strong focus on understanding ecosystems as the fundamental unit of organization of life, explained largely in terms of matter cycles and energy flows within and between ecosystems. For instance, the environmental literacy group (http://envlit.educ.msu.edu/) at the

Michigan State University (MSU) has done a prodigious amount of research to support development of curricular material within an overall ecosystem ecology framework. Their research has led to development of learning progressions on energy flow, carbon, and water cycles in socio-ecological systems (Gunckel, Covitt, Salinas, & Anderson, 2012; Jin & Anderson, 2012; Mohan, Chen, & Anderson, 2009). This group is also a major contributor to the secondary science curriculum project "Carbon TIME" (http://carbontime.bscs.org/) that has developed a series of six teaching modules aimed at helping middle and high school students learn basic biological and ecological processes that transform matter and energy in organisms, ecosystems, and global systems (Anderson et al., 2017). Though much of the research and curriculum development in this area may represent human and natural systems as integrally coupled by matter and energy flow, there is nevertheless a clear ontological distinction between the two systems and an implicit belief in nature-social dualism. For instance, Quigley and Allspaw's (2011) developed an online, five-week unit as a resource for science teachers. This curricular material makes a clear division between "the cultural and ecological 'worlds of central Asia' to show how the 'natural' and 'man-made' worlds influence each other" (p. 71).

Ecological sciences in the last few decades have undergone a major paradigm change that has led to the abandonment of many foundational premises that guided ecosystem ecology for much of the last century. For instance, one of the key assumptions used to be that our world could be understood by dividing it into two distinct ontological domains—a natural and a social world. Ecologists now realize that this assumption makes little sense in the Anthropocene Epoch where the majority of the biosphere lies in agricultural and settled anthropogenic biomes (Ellis, Klein Goldewijk, Siebert, Lightman, & Ramankutty, 2010; Schmitz, 2016). Science educators and researchers have now begun to do the necessary groundwork that is needed to rid school science of such dated assumptions and to re-envision science education so that it offers an interpretation of our world that matches with that of ecologists and environmental scientists. Our book aspires to help the science education community achieve this goal. We hope this book will encourage readers to take a critical look at what students are learning about the world in their science classrooms, and to develop ideas that will better prepare them so that they can bring about a more sustainable and socioecologically just world in the age of Anthropocene.

Organization of the Book

The book can be seen as consisting of three major parts. The first part (Chaps. 1 and 2) sets the stage for the empirical studies that are presented in the second part (Chaps. 3, 4, 5, and 6). In the third part (Chap. 7), we summarize our results and propose a conceptual framework for ecology and environmental science-related topics in science education. The first part includes this and the next chapter, in which we begin by surveying the context in which this book is situated and then proceed to discuss "Evolving Views on the Nature of Nature," in which we present the conceptual framework that guided our work. Chapter 2 begins with an exploration of the modern conception of nature and how it has enabled Western societies to study and exploit the natural world for their own utilitarian purposes since the beginning of the industrial revolution. We then examine the understanding of nature in modern science and its eventual failure to explain a world that is populated by hybrid entities that are both social and natural in all their manifestations and relations. This is followed by an exploration of the emerging contours of an amodern view of nature that currently guides much of research in ecology and environmental sciences. In the end we present the theoretical framework that shaped our research.

The second part of this volume comprises Chaps. 3, 4, 5 and 6 and presents the empirical wherewithal that support our emerging themes and proposed conceptual framework. In the third chapter, "The Intended Curriculum: Locating Nature in the Science Standards," we examine the *Next Generation Science Standards* (NGSS) and the two sets of Georgia science standards—the *Georgia Performance Standards* (GPS) that were used till the end of the academic year 2016–17 and the newly adopted *Georgia Standards for Excellence* (GSE) that were made operational for Georgia public schools from the academic year 2017–18. We show how the intended curriculum as reflected in these standards represents the natural world as a biophysical system that can be "terraformed" and sustainably managed by science and technology to support "green" capitalist societies on this planet. The fourth chapter, "The Intended Curriculum: Nature as Represented in a Science Textbook," is a reprint of our paper (Sharma & Buxton, 2015) that explores how the language of science textbooks works to represent the world for students in distinct ways that have serious implications for their ecological literacy. Our results show that science textbooks may be offering outdated representations of natural systems' relationships with social systems and the role of human agency in these relationships. In the fifth chapter, "The Enacted Curriculum:

Representations of Nature in Science Teaching," we present a case study of how one teacher represented nature while teaching seventh-grade science in her classroom. Here we see how the teacher taught in ways that largely naturalized the normative interpretation of the world as inscribed in the intended curriculum. Finally, in the sixth chapter, "The Received Curriculum: Nature as Understood by Students," we explore the curriculum as reflected in students' understanding of the natural world and its relationship with the human world. On the whole, we see a huge transmission loss between the intended curriculum, the enacted curriculum, and the curriculum that is received, or learned by the students, though this transmission loss is largely limited to the understanding of science concepts. The overall perspectives on the world as embedded in the scientific and environmental discourses of the science standards, on the other hand, manage to seep into the received curriculum with remarkable fidelity.

The seventh and final chapter, "A Sustainability Science based Framework for Science Education" constitutes the third part of the book. Here, after summarizing the main results of our empirical investigations, we present our proposal for how science education should represent ecology and environmental science topics for the needs of the Anthropocene Epoch. Our sustainability science-based framework for science education is defined by (a) acknowledgment of an explicit ethical stance that needs to be an integral part of science education, (b) its compatibility in terms of content with the latest developments in ecology and environmental sciences, and (c) a definite orientation towards praxis in the service of social-ecological justice.

NOTE

1. One of us (Sharma) was part of this group in the initial stages of the research project.

REFERENCES

Aguayo, C., Higgins, B., Field, E., Nicholls, J., Pudin, S., Tiu, S. A., ... Mah, J. (2016). Perspectives from emerging researchers: What next in EE/SE research? *Australian Journal of Environmental Education, 32*(1), 17–29. https://doi.org/10.1017/aee.2015.57

Anderson, C. W., de los Santos, E. X., Bodbyl Roels, S., Covitt, B., Edwards, K. D., Hancock, J. B., ... Welch, M. (2017). *Designing educational systems to support enactment of the Next Generation Science Standards.* CarbonTIME. Retrieved from http://carbontime.bscs.org/conference-presentations

Anderson, M. (2017). *For Earth Day, here's how Americans view environmental issues*. Pew Research Center. Retrieved from http://www.pewresearch.org/fact-tank/2017/04/20/for-earth-day-heres-how-americans-view-environmental-issues/

Assaraf, O. B. Z., & Damri, S. (2009). University science graduates' environmental perceptions regarding industry. *Journal of Science Education and Technology, 18*(5), 367–381.

Bäckstrand, K., & Lövbrand, E. (2006). Planting trees to mitigate climate change: Contested discourses of ecological modernization, green governmentality and civic environmentalism. *Global Environmental Politics, 6*(1), 50–75.

Barnosky, A. D., Matzke, N., Tomiya, S., Wogan, G. O. U., Swartz, B., Quental, T. B., ... Ferrer, E. A. (2011). Has the Earth's sixth mass extinction already arrived? *Nature, 471*(7336), 51–57.

Boysen, L. R., Lucht, W., Gerten, D., Heck, V., Lenton, T. M., & Schellnhuber, H. J. (2017). The limits to global-warming mitigation by terrestrial carbon removal. *Earth's Future, 5*(5), 463–474. https://doi.org/10.1002/2016EF000469

Brulle, R. J. (2000). *Agency, democracy, and nature: The U.S. environmental movement from a critical theory perspective*. Cambridge, MA: MIT Press.

Cheng, J. C., & Monroe, M. C. (2012). Connection to nature: Children's affective attitude toward nature. *Environment and Behavior, 44*(1), 31–49. https://doi.org/10.1177/0013916510385082

Clark, B., & York, R. (2005). Carbon metabolism: Global capitalism, climate change, and the biospheric rift. *Theory and Society, 34*(4), 391–428.

Cobern, W. W. (2000). *Everyday thoughts about nature: A worldview investigation of important concepts students use to make sense of nature with specific attention to science*. Dordrecht, The Netherlands: Springer.

Colucci-Gray, L., Camino, E., Barbiero, G., & Gray, D. (2006). From scientific literacy to sustainability literacy: An ecological framework for education. *Science Education, 90*(2), 227–252.

Covitt, B. A., Tan, E., Tsurusaki, B. K., & Anderson, C. W. (2009). *Students' use of scientific knowledge and practices when making decisions in citizens' roles*. Paper presented at the annual conference of the National Association for Research in Science Teaching. Garden Grove, CA. http://edr1.educ.msu.edu/EnvironmentalLit/publicsite/html/report_2009.html.

Crowell, A., & Schunn, C. (2014). Scientifically literate action: Key barriers and facilitators across context and content. *Public Understanding of Science, 23*(6), 718–733. https://doi.org/10.1177/0963662512469780

Diamond, J. (2013). *Collapse: How societies choose to fail or survive*. London, UK: Penguin Books Limited.

Dietz, T., Fitzgerald, A., & Shwom, R. (2005). Environmental values. *Annual Review of Environment and Resources, 30*(1), 335–372. https://doi.org/10.1146/annurev.energy.30.050504.144444

Dietz, T., Ostrom, E., & Stern, P. C. (2003). The struggle to govern the commons. *Science, 302*(5652), 1907–1912.
Dryzek, J. S. (2013). *The politics of the Earth: Environmental discourses*. Oxford, UK: Oxford University Press.
Eagles, P. F. J., & Demare, R. (1999). Factors influencing children's environmental attitudes. *The Journal of Environmental Education, 30*(4), 33–37. https://doi.org/10.1080/00958969909601882
Ehrlich, P. R., & Ehrlich, A. H. (2013). Can a collapse of global civilization be avoided? *Proceedings of the Royal Society B: Biological Sciences, 280*(1754), 20122845.
Ellis, E. C., Klein Goldewijk, K., Siebert, S., Lightman, D., & Ramankutty, N. (2010). Anthropogenic transformation of the biomes, 1700 to 2000. *Global Ecology and Biogeography, 19*(5), 589–606. https://doi.org/10.1111/j.1466-8238.2010.00540.x
Ergazaki, M., & Ampatzidis, G. (2012). Students' reasoning about the future of disturbed or protected ecosystems and the idea of the "balance of nature". *Research in Science Education, 42*(3), 511–530.
Feinstein, N. W., & Kirchgasler, K. L. (2015). Sustainability in science education? How the next generation science standards approach sustainability, and why it matters. *Science Education, 99*(1), 121–144. https://doi.org/10.1002/sce.21137
Funk, C., & Kennedy, B. (2016). *Public views on climate change and climate scientists*. Washington, DC: Pew Research Center.
Gruenewald, D. A. (2004). A Foucauldian analysis of environmental education: Toward the socioecological challenge of the Earth charter. *Curriculum Inquiry, 34*(1), 71–107. https://doi.org/10.1111/j.1467-873X.2004.00281.x
Gunckel, K. L., Mohan, L., Covitt, B. A., & Anderson, C. W. (2012). Addressing challenges in developing learning progressions for environmental science literacy. In A. C. Alonzo & A. W. Gotwals (Eds.), *Learning progressions in science* (pp. 39–75). Rotterdam, The Netherlands: Sense Publishers.
Hansen, J., Sato, M., Kharecha, P., Beerling, D., Berner, R., Masson-Delmotte, V., ... Zachos, J. C. (2008). Target atmospheric CO2: Where should humanity aim? *Open Atmospheric Science Journal, 2*, 217–231.
Hansson, L. (2014). Students' views concerning worldview presuppositions underpinning science: Is the world really ordered, uniform, and comprehensible? *Science Education, 98*(5), 743–765.
Hardin, G. (2009). The tragedy of the commons. *Journal of Natural Resources Policy Research, 1*(3), 243–253.
Harper, C., & Snowden, M. (2017). *Environment and society: Human perspectives on environmental issues*. New York, NY: Taylor & Francis.
Hempel, L. C. (1996). *Environmental governance: The global challenge*. Washington, DC: Island Press.

Hovardas, T., & Korfiatis, K. (2011). Towards a critical re-appraisal of ecology education: Scheduling an educational intervention to revisit the 'balance of nature' metaphor. *Science & Education, 20*(10), 1039–1053. https://doi.org/10.1007/s11191-010-9325-0

Hufnagel, E., Kelly, G. J., & Henderson, J. A. (2017). How the environment is positioned in the next generation science standards: A critical discourse analysis. *Environmental Education Research*, 1–23. https://doi.org/10.1080/13504622.2017.1334876

Jenkins, E. W. (2003). Environmental education and the public understanding of science. *Frontiers in Ecology and the Environment, 1*(8), 437–443.

Jin, H., & Anderson, C. W. (2012). A learning progression for energy in socio-ecological systems. *Journal of Research in Science Teaching, 49*(9), 1149–1180.

Kahn, B. (2017). We just breached the 410 ppm threshold for CO2. *Scientific American*. Retrieved from https://www.scientificamerican.com/article/we-just-breached-the-410-ppm-threshold-for-co2/

Kempton, W., Boster, J. S., & Hartley, J. A. (1995). *Environmental values in American culture*. Cambridge, MA: MIT Press.

Lefkaditou, A., Korfiatis, K., & Hovardas, T. (2014). Contextualizing the teaching and learning of ecology: Historical and philosophical considerations. In M. R. Matthews (Ed.), *International handbook of research in history, philosophy and science teaching* (pp. 523–550). Dordrecht, The Netherlands: Springer.

Li, J., & Ernst, J. (2015). Exploring value orientations toward the human–nature relationship: A comparison of urban youth in Minnesota, USA and Guangdong, China. *Environmental Education Research, 21*(4), 556–585. https://doi.org/10.1080/13504622.2014.910499

Lindahl, M. G., & Linder, C. (2015). What's natural about nature? Deceptive concepts in socio-scientific decision-making. *European Journal of Science and Mathematics Education, 3*(3), 250–264.

Liu, S., & Lin, H. (2014). Undergraduate students' science-related ideas as embedded in their environmental worldviews. *International Journal of Science and Mathematics Education, 12*(5), 1001–1021.

McGrath, M. (2017). *Donald Trump forest climate change project gains momentum*. Retrieved from http://www.bbc.com/news/science-environment-40927667

Melnick, D. J., Pearl, M. C., & Warfield, J. (2015). Make forests pay. *The New York Times*. Retrieved from https://www.nytimes.com/2015/01/20/opinion/a-carbon-offset-market-for-trees.html

Miller, J. D. (2016). *Civic scientific literacy in the United States in 2016*. Retrieved from http://home.isr.umich.edu/files/2016/10/NASA-CSL-in-2016-Report.pdf

Mohan, L., Chen, J., & Anderson, C. W. (2009). Developing a multi-year learning progression for carbon cycling in socio-ecological systems. *Journal of Research in Science Teaching, 46*(6), 675–698.

Monastersky, R. (2009). Climate crunch: A burden beyond bearing. *Nature, 458*(7242), 1091–1094.
Mueller, M. P., & Tippins, D. J. (2012). Citizen science, ecojustice, and science education: Rethinking an education from nowhere. In *Second international handbook of science education* (pp. 865–882). Dordrecht, The Netherlands: Springer.
National Science Board. (2016). *Science and engineering indicators 2016.* Alexandria, VA: National Science Foundation. Retrieved from https://www.nsf.gov/statistics/2016/nsb20161/#/
Nielsen, J. A. (2012). Arguing from nature: The role of 'nature' in students' argumentations on a socio-scientific issue. *International Journal of Science Education, 34*(5), 723–744.
Payne, P., Cutter-Mackenzie, A., Gough, A., Gough, N., & Whitehouse, H. (2014). Children's conceptions of nature. *Australian Journal of Environmental Education, 30*(1), 68.
Pearce, F. (2006). *When the rivers run dry: Water – The defining crisis of the twenty-first century.* Boston, MA: Beacon Press.
Pedretti, E., & Nazir, J. (2011). Currents in STSE education: Mapping a complex field, 40 years on. *Science Education, 95*(4), 601–626. https://doi.org/10.1002/sce.20435
Pointon, P. (2014). 'The city snuffs out nature': Young people's conceptions of and relationship with nature. *Environmental Education Research, 20*(6), 776–794. https://doi.org/10.1080/13504622.2013.833595
Quigley, C., & Allspaw, K. (2011). The cultural and ecological "worlds" of Central Asia. *Science Scope, 35*(2), 71–74.
Raftery, A. E., Zimmer, A., Frierson, D. M. W., Startz, R., & Liu, P. (2017). Less than 2 °C warming by 2100 unlikely. *Nature Climate Change, 7,* 637. doi:https://doi.org/10.1038/nclimate3352. https://www.nature.com/articles/nclimate3352#supplementary-information
Ranney, M. A., & Clark, D. (2016). Climate change conceptual change: Scientific information can transform attitudes. *Topics in Cognitive Science, 8*(1), 49–75.
Ruppert, J., & Duncan, R. G. (2017). Defining and characterizing ecosystem services for education: A Delphi study. *Journal of Research in Science Teaching, 54*(6), 737–763. https://doi.org/10.1002/tea.21384
Sadler, T. D., & Dawson, V. (2012). Socio-scientific issues in science education: Contexts for the promotion of key learning outcomes. In *Second international handbook of science education* (pp. 799–809). Dordrecht, The Netherlands: Springer.
Schmitz, O. J. (2016). *The new ecology: Rethinking a science for the Anthropocene.* Princeton, NJ: Princeton University Press.

Sellers, C. C. (2012). *Crabgrass crucible: Suburban nature and the rise of environmentalism in twentieth-century America*. Chapel Hill, NC: University of North Carolina Press.

Sharma, A., & Buxton, C. (2015). Human-nature relationships in school science discourse: A critical discourse analysis of a middle grade science textbook. *Science Education, 99*(2), 260–281.

Shepardson, D. P., Roychoudhury, A., & Hirsch, A. S. (Eds.). (2017). *Teaching and learning about climate change: A framework for educators*. New York, NY: Taylor & Francis.

Shepherd, J., Iglesias-Rodriguez, D., & Yool, A. (2007). Geo-engineering might cause, not cure, problems. *Nature, 449*(7164), 781–781.

Steffen, W., Crutzen, P. J., & McNeill, J. R. (2007). The Anthropocene: Are humans now overwhelming the great forces of nature. *Ambio: A Journal of the Human Environment, 36*(8), 614–621.

Stephen, M. G. (2016). Geoengineering: Ethical questions for deliberate climate manipulators. In S. M. Gardiner & A. Thompson (Eds.), *The Oxford handbook of environmental ethics*. Oxford, UK: Oxford University Press.

Sterman, J. D., & Sweeney, L. B. (2007). Understanding public complacency about climate change: Adults' mental models of climate change violate conservation of matter. *Climatic Change, 80*(3), 213–238. https://doi.org/10.1007/s10584-006-9107-5

Tikka, P. M., Kuitunen, M. T., & Tynys, S. M. (2000). Effects of educational background on students' attitudes, activity levels, and knowledge concerning the environment. *The Journal of Environmental Education, 31*(3), 12–19. https://doi.org/10.1080/00958960009598640

Treanor, B. (2010). Environmentalism and public virtue. *Journal of Agricultural and Environmental Ethics, 23*(1), 9–28.

Tsurusaki, B. K., & Anderson, C. W. (2010). Students' understanding of connections between human engineered and natural environmental systems. *International Journal of Environmental and Science Education, 5*(4), 407–433.

Turin, D. R. (2014). Environmental problems and American politics: Why is protecting the environment so difficult? *Inquiries Journal/Student Pulse, 6*(11). Retrieved from http://www.inquiriesjournal.com/a?id=943

Vining, J., Merrick, M. S., & Price, E. A. (2008). The distinction between humans and nature: Human perceptions of connectedness to nature and elements of the natural and unnatural. *Human Ecology Review, 15*(1), 1.

Weaver, A. A. (2002). Determinants of environmental attitudes. *International Journal of Sociology, 32*(1), 77.

Zimmerman, C., & Cuddington, K. (2007). Ambiguous, circular and polysemous: Students' definitions of the "balance of nature" metaphor. *Public Understanding of Science, 16*(4), 393–406. https://doi.org/10.1177/0963662505063022

CHAPTER 2

Evolving Views on the Nature of Nature

"Nature" is a much-used phrase in both everyday and academic language. For instance, a comparative analysis of this word on the online phrase usage graphing tool, Google's "Ngram Viewer," that lets users track the frequency of usage of any word in Google's text corpora in several languages, shows that as a keyword it outranks two commonly used phrases in academia—*culture* and *society*—across 5 million books and over 200 years. It does not surprise us then, that over time, it has not only become "perhaps the most complex word in the language" (Williams, 1985; p. 219), it also has acquired multiple interpretations to become a sort of floating signifier "whose meaning can be gleaned only by relating it to other more directly recognizable signifiers" such as biodiversity and climate change (Swyngedouw, 2015; p. 133). People use this term to refer to (a) the "intrinsic nature," the essential quality of character of an object; (b) the "universal nature," the force that directs the world, such as in "natural laws" or "mother nature"; or (c) the "external nature," the external material world that may or may not include human beings (Ginn & Demeritt, 2008; Williams, 1985). In the Western world, "nature" has also come to acquire a normative valence by alluding to our collective cultural fascination about an ideal world that is environmentally pure and socially perfect and by representing a "norm against which deviations are measured" (Morton, 2007; p. 14).

© The Author(s) 2018
A. Sharma, C. Buxton, *The Natural World and Science Education in the United States*, https://doi.org/10.1007/978-3-319-76186-2_2

From this heteroglossic backdrop of diverse interpretations, and in keeping with the overarching theme of our book, we have chosen one specific articulation of nature prevalent in science education that connects it to "natural systems" or the "natural world" understood as the material but living biophysical world on our planet. This articulation is common in curricular statements like:

(a) "Evaluate or refine a technological solution that reduces impacts of human activities on natural systems." (HS-ESS3-4: Next Generation Science Standards) (Next Generation Science Standards Lead States, 2013)
(b) "Science is both a body of knowledge that represents a current understanding of natural systems and the processes used to refine, elaborate, revise, and extend this knowledge." (Understandings about the Nature of Science: Next Generation Science Standards) (Next Generation Science Standards Lead States, 2013)
(c) Systems, Order, and Organization: "The natural and designed world is complex; it is too large and complicated to investigate and comprehend all at once." (National Science Education Standards) (National Research Council, 1996)

The natural world or the natural systems are indeed fundamental to the knowledge of the world that schools expect students to understand through school science. Predictably, what students learn about the natural world in science classrooms also contributes to their orientation towards the world more broadly and the profound ecological issues confronting it.

In this book we use research done by us and our peers to explore this critical component of current science education in the US context. However, given the remarkable diversity of perspectives on nature even among researchers, it is apposite that before we venture ahead, we adopt a conceptual vantage point that enables us to interpret the current research on this issue as well as propose an alternative theoretical framework that is scientifically up to date as well as sociopolitically progressive.

We begin this chapter by exploring how the modern age led Western societies to understand nature in ways that enabled them to study and exploit the biophysical world for their own purposes. This discursively produced and materially focused understanding of nature influences not only how the public thinks about nature, but remarkably, also continues to shape scholarship in some areas of scientific inquiry. The modern perspective on nature, as this book argues, also seeps into school science through curricula, textbooks, and classroom discourse.

However, the drawbacks of such a discursive construction of nature have become apparent to the scientific community at large, and several disciplines have begun to respond by proposing alternate perspectives on humans and their relationship with the world. We outline some key alternative frameworks that now shape research and scholarly conversations on nature and our relationship with it. We end the chapter by drawing upon these alternative perspectives to delineate our theoretical framework that guided our research presented in the remainder of the book.

CONCEPTIONS OF NATURE AND THE RISE OF THE MODERN AGE

Our current perceptions and attitudes about nature are complex, multihued, and multilayered quite like our relationship with it (Gifford & Sussman, 2012). However, underlying all the complexities are a few key ideas that have been critical to Western civilization's relationship with nature. In this section, we elaborate upon these ideas with a view to understanding their influence on science broadly, and on school science particularly. We begin by tracing the development of these ideas over time in Europe. It is clear from detailed and recent accounts, such as Coates (2013) and Hanawalt and Kiser (2008), that premodern European societies harbored historically contingent, situated, complex, and often contradictory attitudes towards the nonhuman or natural world. These attitudes ranged from highly adversarial and gloomy to collaborative and optimistic, conservationist to exploitative and from inquiry oriented and recreational to magical and suspicious. Further, different facets of nature seemed to occasion diverse set of attitudes.

Despite the broad range of attitudes, we can still discern a few common threads coursing through this multiplicity. First, though the perceived nature of people's relationship with the natural world evolved throughout the Middle Ages, it is evident from the writings of that period that facets of the natural world were seen as distinct from the social world. Texts of that period portray nature or its constituents anthropomorphically as powerful deities, such as Goddess Natura or the Mother Earth, with whom humans had a complex relationship that could be contentious and violent or collaborative and peaceful depending upon the social and environmental circumstances (Aberth, 2012; Economou, 2002). Despite this distinction, people perceived nature to be everywhere and constitutive of all the important contexts of their lives. They "marked personal, community, daily, and seasonal events by natural occurrences and built their cultural explanations around the workings of nature, which formed the unspoken backdrop for every historical event and document of the time" (Hanawalt & Kiser, 2008; p. 1). That is, nature, though distinct from the social, was nonetheless intricately and intimately imbricated with every social, cultural, and religious aspect of people's lives.

Second, given that the natural world, personified as a deity, was seen as agentive and willful, human relationship with the natural world was culturally constrained by ethical and moral limits that if transgressed, could invite ecological retribution from nature. Mother Earth could be a nurturing mother sustaining humanity with its bounty, while also being capable of reprisal against those who harmed her. For instance, death and destruction caused by the Black Death in the Middle Ages was understood to be nature's revenge for the harm done by humans to the natural world (Aberth, 2012). Such assumptions imposed social and moral limits on the extent to which the Earth could be exploited for individual or collective gains (Merchant, 2013). Third, from the time of the ancient Greeks, Western thought was marked by a strong belief in the balance of nature in the world (Kricher, 2009). According to this idea, natural systems have a strong tendency to be in a state of stable equilibrium (homeostasis), and any small disturbance, such as a change in the population of a particular species, is counteracted by the system to restore the original balance between different components of the system. In premodern Europe it was believed that the balance of nature along with order and harmony in the world was ordained and sustained by God. Of course, the belief in balance of nature has also been found to be an integral part of most cosmologies all over the world (Egerton, 1973).

With the arrival of the modern age, following the late Middle Ages in Europe, humans' relationship with the natural world underwent a profound transformation. Modernity marked the age of rationalism in which scientists and philosophers sought to build the edifice of knowledge on the solid foundations of reason and empirical observation. As Koch (1993) argues, in the modern era it was believed that reason, defined as the "objective" power of the mind, would uncover the universal principles that govern the world. Reason would free humankind from ignorance and superstition. Armored with a newly acquired "will to truth," intellectuals of the era became confident enough to build a stable view of the world that rested on a belief in external, objective reality that could be assessed through reason and scientific methods. Further, along with these transformational changes in Western thought, another monumental shift took place in modern societies with regard to their material relationship with the world around them. The keystone of the human relationship with the natural world has always been a continuous exchange of matter and energy between humans and their natural world. Marx labeled this exchange as social metabolism as it resembles the unceasing exchange of matter and energy that all living beings have with their surroundings (Foster, 1999). In premodern Europe, like all traditional societies around the world, people sustained themselves materially by drawing on nearby resources with which they had direct contact. Their food came from neighboring farms and their wood from surrounding forest lands. However, with the onset of the industrial revolution that accompanied the modern age, people began to draw on distant sources for their material sustenance and consequently became increasingly estranged from the natural world surrounding them. That is, as first indicated, the industrial revolution in Western Europe created a metabolic rift between humans and their surrounding natural world with which they had hitherto exchanged matter and energy (Foster, 1999).

As a result of the material and conceptual changes brought about by the rise of modernity, humans began to reposition their relationship with the world. At an intellectual level, modern philosophy likewise began to grapple with the problem of understanding the cleavages and contradictions between the ways, values, and institutions of the old premodern world and the new modern developments that were shaping science, industry, and the arts. Acutely aware of these challenges, European intellectuals sought to reconcile the new with the old by formulating these divisions as dualisms. According to Dewey (2008), these dualisms soon became "the staple of modern philosophy" (p. 407) and are "so conspicuous that they come readily to mind: The material and spiritual, the physical and the mental or psychical, body and mind, experience and reason; sense and intellect, appetitive desire and will; subjective and objective; individual and social; inner and outer" (p. 408). With regard to the natural world, these modern societies began to objectify nature and set themselves apart from the rest of the world using a consolidated worldview, labeled as the *Modern Constitution* by Latour (2012). This view ontologically divided the world into two distinct and separate realms—nature and society —by inventing "a separation between the scientific power charged with representing things and the political power charged with representing subjects" (p. 29). Natural sciences developed to uncover the laws of nature, and social sciences blossomed to arrive at the precise knowledge of how our societies functioned. In this separation between "science of things" and the "politics of men," the Modern Constitution rendered invisible and unrepresentable the diverse hybrid entities and processes that consisted of both natural and social components and thus could not be accommodated within the nature-social dualism. However, this lack of representation did not limit in any way the mediation that these hybrid entities accomplished between the two ontological spheres of nature and society. For instance, any modern technological invention designed to work on objects of nature must be a nature-society hybrid, but guided by the Modern Constitution, humans in Western societies assumed that such inventions would not have any impact on the social world as the two ontological domains—the nature and the society—were expected to be governed by their own distinct laws.

Further, with the arrival of the modern age, it is not surprising that belief in a deity-like nature with whom humans could have an ethical relationship began to feel like ignorant, superstitious folklore. An ethical and moral relationship in which humans could propitiate or anger an animistic, anthropomorphic nature gradually gave way to a mechanical view that saw nature "as a system of dead, inert particles moved by external, rather than inherent forces" which could be dominated and manipulated by humans, who were now armed with scientific knowledge and technology (Merchant, 2013). No longer encumbered by moral and ethical restraints, humans, therefore, began to exploit the natural world, guided by an unrestrained instrumental rationality and the imperatives of capitalism. This new relationship with nature was accompanied and aided by a rising belief in a set of assumptions about the place of the human beings in the natural order. These assumptions, labeled as the *Human Exceptionalism Paradigm*, asserted that "Humans have a cultural heritage in addition to (and distinct from) their genetic inheritance, and thus are quite unlike all other animal species" (Bowden, 2004), and thus they could exercise domination over all other species on the planet. Further, in contrast to the belief systems of the medieval ages in which nature was critical to all human affairs, the human exceptionalism paradigm elevated social and cultural environments as the key determinants of our social and individual lives, while relegating the natural world to the periphery.

As the unsustainable exploitation and commodification of nature reveals, the human relationship with the natural world, at a socioeconomic level, is still governed by this human exceptionalism paradigm

(Manuel-Navarrete & Buzinde, 2010). However, at a personal level many people in the West are now beginning to shift their personal beliefs to a *New Ecological Paradigm*, which positions them more as stewards or guardians of the natural world and as interdependent with other species, rather than as masters of the natural world (Braito, Böck, Flint, Muhar, Muhar, & Penker, 2017; Corner, Parkhill, Pidgeon, & Vaughan, 2013; de Groot, Drenthen, & de Groot, 2011). Interestingly, amidst these changes, the belief in balance of nature overseen by supernatural forces has persisted and still constitutes an important part of the environmental belief systems in Western societies (Bechtel, Verdugo, & de Queiroz, 1999; Kricher, 2009). Though Darwin and many other scientists have tried to supplant God with evolution as the basis for the balance of nature, the persistence of pseudoscientific creationist beliefs in the West indicates a striking endurance of premodern beliefs about nature in these societies (Newport, 2014; Simberloff, 2014).

Nature and Modern Science

Evolving societal beliefs about nature in the modern world also powerfully shaped the epistemological and ontological foundations of modern science. For instance, as Lewontin and Levins (2007) assert, the nature-social dualism molded the discipline of biology such that biologists began to look at the world as one where humans were positioned as the external force that disturbs the natural state of harmony and equilibrium in nature. The key task for these biologists then became "to uncover the laws of behavior of the unperturbed natural world and to use these laws to hold in check the effects of the external perturbing force" (p. 16). Ecologists too preferred to "do their studies in wild places devoid of human influence, because it was held that ecological and evolutionary processes were not anthropogenic in origin" (Schmitz, 2016; p. 8). This dualism, created a "great divide" between the social and natural sciences by splitting the epistemological domains of the natural and social sciences (Hausknost et al., 2016) and by presenting "nature as the externally conditioning frame for the human life – an externalization that permitted the social sciences and humanities to leave the matter of nature to the natural sciences" (Swyngedouw, 2015; p. 132). Even today, much research in the fields of physical geography (Castree, 2005), ecofeminism (Nhanenge, 2011), and social ecology (Stone-Jovicich, 2015) carries the underlying foundational assumption of nature and social as two distinct ontological domains.

In addition, steered by the modernist impulse to discover universal laws that reveal the objective truths about the world and the desire to emulate the success of physics in explaining the material world, biological sciences, especially ecology, has striven to become a nomothetic science "whose highest goal was to produce broad, context-independent generalizations about nature" (Demeritt, 1994; p. 23). The premodern belief in the balance of nature has also been hard to shake off in biological sciences (Simberloff, 2014). Guided by this ancient belief, ecologists for much of the twentieth century viewed ecosystems as having "a strategy of self-regulation replete with an integrated and homeostatic system, governed by their own organic laws and ability to respond to feedbacks in accordance with the mechanistic principles of cause and effect all directed towards achieving internal equilibrium or balance" (Jelinski, 2010; pp. 41–42). This belief was also an important component of the worldview of conservation movements all over the world (Jelinski, 2005). In fact, when Ladle and Gillson (2009) investigated whether the "balance of nature" metaphor is being shelved in favor of the more scientifically up-to-date and dynamic "flux of nature" metaphor in the conservation and ecosystem management literature in news media, the internet, and academic literature, they found that "the media and the global Internet community still portray the aim of conservation science and of conservationists as being one of maintaining stability, harmony and balance" (p. 229).

In the last few decades of the twentieth century, the foundational beliefs in the modern view of nature came under increasing assault by both research and world events. For example, the increasing commodification of nature has been accompanied by the proliferation of hybrid entities, objects, and processes, such as climate change, genetically modified or cloned animals, and the ozone hole, each of which emerges out of a complex mixture of natural, social, cultural, and political aspects of our world, thus challenging our belief in a social-nature dualism. Under the Modern Constitution these entities could neither be satisfactorily categorized as social or nature and thus were largely kept out of our discursive frame despite powerfully shaping social and ecological aspects of our lives. These "illegitimate couplings," "monsters," "quasi-objects," or "naturecultures," as Latour (2012) and Haraway (2008) have called them, have now become too numerous and influential to be excluded from scientists' ontological commitments about the world. Further, as Bradshaw and Bekoff (2001) admit, the intensification of environmental problems in the twentieth century also forced ecologists to realize that ecology can only succeed at helping them understand the world once they do away with the century-old separation of the natural and the social world. As a result, the last few decades have seen a profound shift in scientific understanding of our place and role in the world. Though many scientists still use the social-nature dualism as an analytic device to help interpret their research, the ontological division of the world in two distinct domains—nature and social—is now largely considered outdated and unhelpful by the majority of scientists (Schmitz, 2016). This emergent perspective has not, however, become widespread among environmental activists and the general public, where the nature-social dualism still shapes their ontological outlook on the world (Braun & Castree, 2005).

Further, the hopes that ecology and other biological sciences will be able to emerge as a serious nomothetic science *à la* Physics have largely been doused by the complex and chaotic nature of our world that has successfully resisted being reductively described by mathematical models and universal laws (Lewontin & Levins, 2007). For much of the twentieth century, systems ecologists tried to understand the biophysical world in terms of cybernetic systems in which different "black boxed" entities represent unexamined complex and diverse dynamical relationships between organisms that were engaged in mutual fluxes of matter of energy. But now we know that this ecosystem-based paradigm is a rather poor match for the complexity of the phenomena it sought to explain (Pickett, 2013; Schizas, 2012).

There is also an increasing realization among ecologists that the foundational definitions and concepts of ecology such as ecosystem, niche, community, and population are based on simplistic assumptions about the world (Demeritt, 1994). After more than a century of research, serious questions remain regarding the definition of these terms, such as "whether they are defined statistically or via a network of interactions", and "if their boundaries are drawn by topographical or process-related criteria" (Jax, 2006; p. 237). The resulting vagueness in these critical concepts is seen as a serious challenge to the construction of coherent and powerful theories in ecology (Sagoff, 2003). It should not be surprising then that despite more than a century of hard work, biological sciences have little to show by way of broad, context-independent generalizations about nature.

Lastly, over the last three decades, it also became obvious to ecologists that equilibrium conditions are rare and "disturbance events so common that most ecological systems never reach equilibrium, including vegetation even over large landscapes" (Jelinski, 2010; p. 44). As a result, the idea that natural systems self-regulate themselves to maintain or achieve a state of balance or equilibrium is no longer taken seriously by most ecologists (Capinera, 2008). However, like other modern beliefs about nature, this belief continues to run strong among natural resource management experts, landowners, policymakers, media reports, and the general public (Hull, Robertson, Richert, Seekemp, & Buhyoff, 2002; Ladle & Gillson, 2009).

Emerging Contours of an Amodern View of Nature

Faced with the challenges outlined in the previous section, the scientific community has responded by making substantial revisions in their understanding of nature. First, taking inspiration from Marx's thesis of unity of nature with society as produced in practice through human labor, critical human geographers from Harvey (1982) and Smith (1984) onwards have managed to persuade the scientific community to see nature not as something that exists "out there" as an objective reality, but rather, as a social construction that is produced by human material and discursive practices. That is, as Archer (2010) puts it, "Nature is always social nature, in this respect, not something outside social reproduction to be merely observed, protected, saved, restored, conserved, or otherwise managed by somehow extranatural humans. Nature is what results from the various cultural and historical ways in which humans socially (re)produce both their sciences and their societies" (p. 2560). Research in environmental history has shown us the historically contingent ways in which our understanding of nature has changed over time. For instance, there was a time not far back in American and European colonial history when "nature" stood as the antithesis of civilization and things and people and places associated with nature were seen negatively as savage, less civilized, deserted, and barren (Ginn & Demeritt, 2008). By the end of the nineteenth century, however, the so-called areas where one could go to find "nature" or "wilderness," basically areas that could not be cultivated or put to any economic use, began to be viewed favorably and were "set aside as national parks or reserves, where nature was to be preserved in an unspoilt state for future generations to admire" (Ginn & Demeritt, 2008; p. 303). That is, as Cronon (1996), explaining about the historically situated social construction of wilderness, argued:

The more one knows of its peculiar history, the more one realizes that wilderness is not quite what it seems. Far from being the one place on Earth that stands apart from humanity, it is quite profoundly a human creation – indeed, the creation of very particular human cultures at very particular moments in human history. It is not a pristine sanctuary where the last remnant of an untouched, endangered, but still transcendent nature can for at least a little while longer be encountered without the contaminating taint of civilization. Instead, it is a product of that civilization, and could hardly be contaminated by the very stuff of which it is made. (p. 7)

The dissolution of nature as an objective reality has led to the rise of *New Ecology* in ecological sciences in which environment is seen "as both the product of and the setting for human interactions, which link dynamic structural analyses of environmental processes with an appreciation of human agency in environmental transformation, as part of a 'structuration' approach" (Scoones, 1999; p. 479). That is, the new ecological approach attempts to overcome the social-nature dualism by investigating the world through a social-ecological systems framework in which it is assumed that the "human political, cultural, religious, and economic institutions influence how nature works" and "feedbacks from nature can instigate institutional change in a co-dependent way" (Schmitz, 2016; p. 7). There are currently multiple social-ecological systems frameworks available to ecologists that differ in terms of whether the relationship between the social and ecological systems is assumed to be uni- or bidirectional, whether the perspective on the ecological system is anthropocentric or ecocentric, and whether the framework is analysis or action oriented (Binder, Hinkel, Bots, & Pahl-Wostl, 2013).

The paradigmatic shift in scientific understanding of nature and the rise of new ecology also includes moving away from ahistorical and nomothetic systems ecology and its explorations of undifferentiated ecosystems towards historical, ideographic investigations of individuals, species, and populations and their direct context-specific relationships, such as in symbiotic systems (Zimmerer, 1994). This shift has translated into a greater emphasis on "spatial and temporal dynamics developed in detailed and situated analyses of 'people in places,' using, in particular, historical analysis as a way of explaining environmental change across time and space" (Scoones, 1999; p. 479). The change can be clearly seen in the increasing influence of landscape ecology for understanding current ecological issues—an interdisciplinary discipline where landscapes are studied as heterogeneous holistic spaces in which nature and culture co-evolve (Wu, 2011).

Finally, as mentioned earlier, ecological sciences have discarded the balance of nature view that highlighted assumptions of steady state, homeostasis, and stability of ecosystems and treated humans as external entities. In its place, we find an increasing acceptance of an inclusive and non-equilibrium perspective that accepts: "1) the material openness of ecological systems; (2) the role of external regulation; (3) the absence or transience of equilibrium states; (4) the commonness and significance of natural and human-caused disturbances; (5) the multiple pathways of system dynamics, and (6) the pervasive involvement of human actors, both local and distant, in ecosystems" (Pickett, 2013; p. 265).

Changing assumptions about the nature of nature have not only led to epistemic shifts in the biological sciences. In the social sciences as well, we see at least three distinct ways in which nature as a social construct is being understood and analyzed.[1] As summarized by Stone-Jovicich (2015), these three approaches are:

(a) *Materio-spatial world systems analysis.* Rooted in sociology and based on world systems theory, this approach assumes that "the contemporary world is so interconnected that it constitutes a whole interactive system, i.e., the global and local, and everything in between, are intricately tied to each other" (para. 11). Consequently, this approach focuses on undertaking materialist, structuralist analyses of world system-level processes and patterns for understanding long-term and cross-scale society-environment relations and dynamics. Though the approach moves beyond the positivist and modern notion of nature as an objective reality, it does maintain a conceptual separation between the social and biophysical world while integrating ecological considerations within the materio-spatial world systems framework.

(b) *Critical realist political ecology.* Much preferred by human geographers, critical realist political ecology undertakes poststructuralist analyses, albeit within a realist ontology, of socioecological change. Like Materio-spatial world systems analysis, this approach also maintains a conceptual distinction between the social and biophysical world. The main objective of these analyses is to highlight the ecological degradation and social inequities in the world with the assumption that many of these outcomes have been caused or worsened by mainstream scientific explanations that largely reflect the

interests and values of the powerful groups in the society. Thus, knowledge production and its contestation are the primary units of analysis in this approach.
(c) *Actor-network theory*. This approach comes from the science and technology studies community of social scientists. Actor-network theory (ANT) not only deconstructs the conceptual category of "nature" but also its obverse side—the social. According to this approach, "the society–nature dualism illicitly simplifies a world that is much messier than we allow. This world does not divide neatly into two ontological domains but is, rather, characterized by myriad qualitatively different but intimately related phenomena" (Castree, 2005; p. 231). That is, dispensing with the notion of nature and social and rejecting the modernist distinctions between objects and subjects, people and machines, material and imaginary, ANT explores the world with the "amodern" ontological assumption that it consists of "hybrid networks composed of specific human and non-human actants, that are of greater or shorter length, are more or less dense, and 'hold together' for longer or shorter periods of time" (Braun, 2006; p. 202). Predictably, human and nonhuman actors and actor networks constitute the primary units of analysis in ANT.

Confronted with the complexities of understanding life on the planet in an era where humans have become the major driver of ecological change, scientists and social scientists have found it expedient to break out of their narrow disciplinary shells to share the respective strengths and insights of their fields and work together on socioecological issues that clearly transcend all disciplinary boundaries. This unique collaboration has led to the emergence of a new twenty-first-century interdisciplinary academic discipline—sustainability science. This new discipline "transcends the concerns of its foundational disciplines and focuses instead on understanding the complex dynamics that arise from interactions between human and environmental systems" (Clark, 2007; p. 1737) with the intention of understanding how these interactions help us achieve the goals of sustainability in terms of "meeting the needs of present and future generations while substantially reducing poverty and conserving the planet's life support systems" (http://sustainability.pnas.org/page/about). Sustainability science is rapidly gaining widespread recognition and influence as reflected in the beginning of several research journals devoted to the discipline as well as new research programs in universities all over the world. We hope that our book contributes to broadening awareness of new trends in sustainability science among science educators as well. Of course, what practicing scientists do and what students learn in the name of science at school are rarely the same thing. However, as we elaborate in this book, science, when re-contextualized from the lab to the school classroom, preserves many of the key features that marked how scientists thought about and researched nature for much of the twentieth century. Upcoming chapters will explore how school science continues to be guided by a firm belief in broad, context-independent concepts and generalizations about nature, social-nature dualism, and balance of nature.

We began the chapter with a brief historical exploration of how societies of the West understood their natural world. This allowed us to appreciate the modern perspective on nature that deeply influenced how both scientists and common folks perceived nature over several centuries. In the closing decades of the twentieth century the modern perspective was discarded by ecologists as well as by social scientists as they realized that an ontologically distinct, stable, and homeostatic natural world is actually a socially produced "reality" and not an objective truth about the world. Unfortunately, as we show in this book, science education in the United States continues to portray the world from this dated modern perspective that no longer reflects the current understanding about the world in science and social science research. Clearly, we need a new framework for teaching about the natural world in our science classrooms. Deconstruction of an established perspective on nature and reconstruction of an alternative one, however, needs a theoretical standpoint that includes both science and education in its ambit. In the next section, therefore, we lay out our theoretical framework that guides our critique of the extant perspectives as well as construction of an alternate perspective on nature in the context of science education.

A New Theoretical Framework for Rethinking School Science

Our work spans both school science content and how that content is taught. Therefore, the conceptual framework of the book has two dimensions. One dimension conceptualizes the nature of school science, especially on ecological issues that we wish to see in schools. The other dimension relates to how science education shapes students' understanding about the world.

Sustainability Science Matters

Let's begin with school science first. The conceptual framework that guided the work reported in this book is grounded in the increasing realization by the scientific community that we have transitioned from the Holocene Epoch to a new geological epoch in Earth's history, called the Anthropocene. In this new epoch, the human influence on the Earth has become so large that it now rivals all other great geologic forces that have shaped our planet since the very beginning (Steffen, Grinevald, Crutzen, & McNeill, 2011). Humans dominate most of Earth's ecosystems. In fact, there is hardly any local or regional ecosystem that is now free of extensive human influence (Steffen, Crutzen, & McNeill, 2007). Scientists are still figuring out how life on the planet is adapting to the epochal changes to Earth's systems brought about by human impact. However, research to date clearly shows that (a) "The planet Earth as a whole is a crisis ridden self-organizing complex system" and (b) "Mankind is an integrated part and a powerful driver of the Earth's systems-dynamics" (Becker, 2012; p. 39). As we argue in this book, the traditional school science framework, on the contrary, assumes a harmonious, stable natural world from which humans can be externalized. Evidently, this outdated perspective will no longer work in the Anthropocene Era, and nor will a STEM-oriented science focused on transmuting science learning into workforce apprenticeship for the industrial sector. This later view in particular risks pushing all other pedagogic and curricular goals to the periphery. If we really wish to see science education as critical to preparing future citizens to survive and thrive in an ecologically precarious world in ways that are both socioecologically just and democratic, a different framework will be required.

Clearly, we need a framework for science education that is not only compatible with the latest developments in science but also prepares students to understand our world in the age of Anthropocene impacts in ways that contribute to the goals of sustainability. This framework should help students understand the complex and dynamic materiality of the socioecological world they inhabit while also making them aware of the social construction that leads to a plurality of natures in a contingent, historical manner in different societies all over the world.

Such a framework would situate sustainability science as the primary model for school science (Carter, 2008; Sharma, 2012). Naturally, the conceptual framework of our work must also be structured by the principles of sustainability science as re-contextualized for the purposes of science education. Based on the three core research dimensions of sustainability science as summarized in Dedeurwaerdere (2014), these overarching principles are:

(a) Ethics: Students' understanding of science content should be imbued with an ethical stance that celebrates equity; democratic contestation; reconciliation through negotiation; a virtue-based ethics of care, kindness, and compassion; as well as an ethics of intellectual rigor and an ethics of civility.
(b) Content: Science education should help students understand the world as consisting of materially open, nested, and coupled socioecological systems. These socioecological systems are to be understood from the resilience perspective that "emphasizes non-linear dynamics, thresholds, uncertainty and surprise, how periods of gradual change interplay with periods of rapid change and how such dynamics interact across temporal and spatial scales" (Folke, 2006; p. 253).
(c) Praxis: Students should begin to work with local and distant communities to explore and help implement collective solutions to sustainability problems, extending the notion of practice, which is at the heart of current science education reforms, to embrace "reflection and action directed at the structures to be transformed." (Freire, 1970).

Thus, our framework on science content is well aligned with the current understanding among scientists about the roles of humans in the biophysical world. It is also compatible with and supports the curricular goals of Science-Technology-Society (STS) education and Socioscientific Studies Issues-based (SSI) Education. Our framework does diverge from most current STSE and SSI education because we present an alternate framework on ecology topics in science education while STSE and SSI education take the current frameworks as given and work from there to connect science with socioecological and social justice issues. Further, as we will show in the next chapter, our framework also has some significant differences with the science curricula standards being currently used in US schools. Of course, what scientists do and what students learn in their science classes are rarely the same thing. In order to enter the realm of school science, the scientists' science must first be re-contextualized (Sharma & Anderson, 2009). The second dimension of our conceptual framework deals with this process of re-contextualization as it has a strong impact on what students learn about the world through school science.

Discourse (Still) Matters

Owing to the dissatisfaction with modern humanist and structuralist theoretical standpoints, we are currently witnessing an 'ontological turn' in philosophy and social sciences that has led scholars to critique theories on the grounds of their ontological commitments. This critique has been particularly directed at theories or theoretical approaches that embody a substantivist ontology of the world in terms of considering entities primary to relations in the order of priority. The "ontological turn," on the other hand, has favored theories or theoretical frameworks, such as new materialism (Barad, 2003), and actor-network methodologies (Latour, 2005), that view the world from a relational standpoint in the sense that they consider relations primary to entities. Being based partially on ideas and concepts circulating in new materialist and actor-network scholarship, our theoretical perspective likewise hews to a relational ontological standpoint. That is, we see the world as consisting of multiple overlapping networks of relations of humans and nonhumans. Here relations are to be seen as material-discursive practices that "enact" or "perform" an ontology into existence based on what people and objects do once embedded in the relations (Braun, 2008; Mol, 1999).

In Western developed societies such as the United States, the re-contextualization of scientists' science into school science occurs in complex, dynamic, and multiscale networks involving an eclectic mix of public, private, and nongovernmental organizations and actors. These networks include (a) international groups, such as the Organisation for Economic Co-operation and Development (OECD) and the International Association for the Evaluation of Academic Achievement (IEA); (b) national organizations and networks, such as the US Department of Education, STEM Funders Network (http://www.stemecosystems.org/), the American Association for the Advancement of Science (AAAS), and Achieve[2]; (c) state-level organizations, such as the state department of education and teacher advocacy groups; and finally local-level organizations and actors, such as the school boards, teachers, parents, and students. These networks organize "centers of translation where network elements are defined and controlled, and strategies for translation are developed and considered" and engage in translation of "materials, actors, and texts into *inscriptions* that allow influence at a distance" (Crawford, 2005; p. 2). The translation practices understood as practices of negotiation, mobilization, and displacement, such as framing of policy documents, writing of curriculum standards and textbooks, and classroom-based instructional practices, together re-contextualize scientists' science into school science, leading to curricular goals, science content, and pedagogic practices for students in school settings.

Like all other relations that constitute a network, translation practices are both material and discursive in nature (Law, 2009). As Barad (2003) asserts, "The relationship between the material and the discursive is one of mutual entailment. Neither is articulated/articulable in the absence of the other; matter and meaning are mutually articulated" (p. 822). We acknowledge the importance of understanding the materiality of relations in any network. In our work, however, we have largely focused on elucidating the discursive elements of translation practices focused on re-contextualization of scientists' science to school science. This choice was dictated both by our understanding of what is more important to explore in a research project such as ours and our methodological limitations as researchers.

We also chose not to expand the scope of this book to include an understanding what nature *is* or what the social conditions are in which our conceptions about nature get produced. The focus of our work is rather different and limited to exploring how nature gets represented, taught, and understood in science classrooms in the United States. Discourses determine the conditions of possibility of how things and actions can and cannot be defined, articulated, produced, or performed. They are critical to the performances or practices that bring objects, phenomena, and actors to life. Through our research we have tried to understand what is possible for science teachers and students to articulate or not articulate about nature given the discourses that constitute the current translation practices regarding school science. Despite its limited focus, we believe this kind of research remains important because as Latour (1996) once opined "when you bracket out the question of reference and that of the social conditions of productions – that is Nature 'out there' and Society 'up there' – what remains is, in a first approximation, meaning production, or discourse, or, text. ... Instead of being means of communications between human actors and nature, meaning productions became the only important thing to study" (p. 8). Further, our society produces multiple articulations of nature in diverse ways, at different levels and contexts and using a disparate cast of actors (Braun & Castree, 2005). Each articulation of nature vies for hegemony over the others in a political struggle of high stakes because the hegemonic representation gets to determine how we relate to the rest of the world. Thus, according to Braun and Castree (2005) it may be important for researchers to show how "nature" gets constituted discursively so that the "self-evidence" of "nature-as-received" can be disrupted "in order to open space for different constructions less implicated in relations of domination" (p. 33). Hence, the importance of our rather limited focus on understanding the discursive elements in relations within actor networks that are engaged in social production of nature in the context of school education in the United States.

Summing Up

The modern perspective on nature has become so taken for granted that it has congealed into an objective truth about the world. We have grown accustomed to believe that humans are distinct and separate from the rest of the world, and the planet will continue to exist the way it has always been since the rise of human civilizations. Such a view helped usher in the unprecedented levels of prosperity and general well-being of the modern era that humans had never experienced before by enabling us to study and exploit nature with utter abandon. That modern era has now slipped into history and we are now in the epoch of the Anthropocene, in which we are confronted by the potentially calamitous outcomes of the Modern Constitution in the form of global climate change and other grave environmental crises.

Biological and social sciences have responded to the arrival of the Anthropocene Epoch by forging alternative perspectives of the world that deconstruct the social-nature dualism and that see humans as an integral component of a dynamic and complex socioecological world. Rethinking the human-nature dualism has also given rise to a new discipline—sustainability science—that is focused on understanding our world in terms of coupled social-ecological systems, manifesting a clear intent to explore ways to exist sustainably and equitably on this planet now and for the foreseeable future. We are convinced that science educators must join in this work by doing a similar re-envisioning of their conceptual framework for science education. In this chapter we laid the conceptual grounds for how such a reconstruction can begin. Our theoretical framework takes the objectives of sustainability science at its heart and offers a view to conceptualize the content, ethics, and praxis of science education, especially as they relate to socioecological issues. This framework also provides us a way to understand how nature is discursively constructed for science students through curricular resources and classroom discourse. With these conceptual tools in hand, we are now ready to forge ahead in the creative tasks of deconstruction and reconstruction of nature in science education.

NOTES

1. At a broader philosophical level, one can see that the dissolution of nature as a conceptual category and an objective reality in ecology and social sciences aligns well with the intellectual zeitgeist of the postmodernism that saw the deconstruction of all conceptual dualisms in philosophy and dispensing of the modernist idea that language is an unproblematic, transparent, and inert medium to apprehend external reality of the world.
2. An independent, nonpartisan, nonprofit education reform organization created by a group of governors and business leaders to raise academic standards and graduation requirements, improve assessments, and strengthen accountability (http://www.achieve.org/).

REFERENCES

Aberth, J. (2012). *An environmental history of the Middle Ages: the crucible of nature*. New York, NY: Routledge.

Archer, K. (2010). Social construction of nature. In B. Warf (Ed.), *Encyclopedia of geography* (pp. 2558–2561). Thousand Oaks, CA: Sage.

Barad, K. (2003). Posthumanist performativity: Toward an understanding of how matter comes to matter. *Signs, 28*(3), 801–831. https://doi.org/10.1086/345321

Bechtel, R. B., Verdugo, V. C., & de Queiroz Pinheiro, J. (1999). Environmental belief systems: United States, Brazil, and Mexico. *Journal of Cross-Cultural Psychology, 30*(1), 122–128. https://doi.org/10.1177/0022022199030001008

Becker, E. (2012). Social-ecological systems as epistemic objects. In M. Glaser (Ed.), *Human-nature interactions in the Anthropocene: Potentials of social-ecological systems analysis* (pp. 37–59). London, UK: Routledge.

Binder, C. R., Hinkel, J., Bots, P. W. G., & Pahl-Wostl, C. (2013). Comparison of frameworks for analyzing social-ecological systems. *Ecology and Society, 18*(4). https://doi.org/10.5751/ES-05551-180426

Bowden, G. (2004). *From environmental to ecological sociology*. Paper presented at the annual meeting of the Australian Sociology Association. La Trobe University.

Bradshaw, G. A., & Bekoff, M. (2001). Ecology and social responsibility: The re-embodiment of science. *Trends in Ecology & Evolution, 16*(8), 460–465.

Braito, M. T., Böck, K., Flint, C., Muhar, A., Muhar, S., & Penker, M. (2017). Human-nature relationships and linkages to environmental behaviour. *Environmental Values, 26*(3), 365–389.

Braun, B. (2006). Towards a new Earth and a new humanity: Nature, ontology, politics. In N. Castree & D. Gregory (Eds.), *David Harvey: A critical reader* (pp. 191–222). Malden, MA: Blackwell Publishing.

Braun, B. (2008). Environmental issues: Inventive life. *Progress in Human Geography, 32*(5), 667–679.

Braun, B., & Castree, N. (2005). *Remaking reality: Nature at the Millenium*. New York, NY: Taylor & Francis.

Capinera, J. L. (2008). Balance of nature. In J. L. Capinera (Ed.), *Encyclopedia of entomology* (pp. 359–359). Dordrecht, The Netherlands: Springer.

Carter, L. (2008). Sociocultural influences on science education: Innovation for contemporary times. *Science Education, 92*(1), 165–181.

Castree, N. (2005). *Nature*. Abingdon, UK: Routledge.

Clark, W. C. (2007). Sustainability science: A room of its own. *Proceedings of the National Academy of Sciences, 104*(6), 1737.

Coates, P. (2013). *Nature: Western attitudes since ancient times*. Malden, MA: John Wiley & Sons.

Corner, A., Parkhill, K., Pidgeon, N., & Vaughan, N. E. (2013). Messing with nature? Exploring public perceptions of geoengineering in the UK. *Global Environmental Change, 23*(5), 938–947. https://doi.org/10.1016/j.gloenvcha.2013.06.002

Crawford, C. S. (2005). Actor network theory. In G. Ritzer (Ed.), *Encyclopedia of social theory* (pp. 1–3). Thousand Oaks, CA: Sage.

Cronon, W. (1996). The trouble with wilderness: Or, getting back to the wrong nature. *Environmental History*, 7–28.

de Groot, M., Drenthen, M., & de Groot, W. T. (2011). Public visions of the human/nature relationship and their implications for environmental ethics. *Environmental Ethics, 33*(1), 25–44.

Dedeurwaerdere, T. (2014). *Sustainability science for strong sustainability*. Cheltenham, UK: Edward Elgar Publishing.

Demeritt, D. (1994). Ecology, objectivity and critique in writings on nature and human societies. *Journal of Historical Geography, 20*(1), 22.

Dewey, J. (2008). *The later works of John Dewey, volume 16, 1925–1953: 1949–1952, essays, typescripts, and knowing and the known*. Carbondale, IL: Southern Illinois University Press.

Economou, G. D. (2002). *The goddess Natura in medieval literature*. Notre Dame, IN: Notre Dame Press.
Egerton, F. N. (1973). Changing concepts of the balance of nature. *QUARTERLY REVIEW OF BIOLOGY*, 322–350.
Folke, C. (2006). Resilience: The emergence of a perspective for social–ecological systems analyses. *Global Environmental Change*, 16(3), 253–267. https://doi.org/10.1016/j.gloenvcha.2006.04.002
Foster, J. B. (1999). Marx's theory of metabolic rift: Classical foundations for environmental sociology. *American Journal of Sociology*, 105(2), 366.
Freire, P. (1970). *Pedagogy of the oppressed*. New York, NY: Continuum.
Gifford, R., & Sussman, R. (2012). Environmental attitudes. In S. Clayton (Ed.), *The Oxford handbook of environmental and conservation psychology* (pp. 65–80). New York, NY: Oxford University Press.
Ginn, F., & Demeritt, D. (2008). Nature: A contested concept. In N. Clifford, S. Holloway, S. P. Rice, & G. Valentine (Eds.), *Key concepts in geography* (pp. 300–311). London, UK: Sage.
Hanawalt, B. A., & Kiser, L. J. (2008). *Engaging with nature: Essays on the natural world in medieval and early modern Europe*. Notre Dame, IN: University of Notre Dame Press.
Haraway, D. J. (2008). *When species meet* (Vol. 224). Minneapolis, MI: University of Minnesota Press.
Harvey, D. (1982). *The limits to capital*. Oxford, UK: Basil Blackwell.
Hausknost, D., Gaube, V., Haas, W., Smetschka, B., Lutz, J., Singh, S. J., & Schmid, M. (2016). 'Society can't move so much as a chair!'—Systems, structures and actors in social ecology. In *Social ecology* (pp. 125–147). Cham, Switzerland: Springer.
Hull, R. B., Robertson, D. P., Richert, D., Seekamp, E., & Buhyoff, G. J. (2002). Assumptions about ecological scale and nature knowing best hiding in environmental decisions. *Conservation Ecology*, 6(2), 12.
Jax, K. (2006). Ecological units: Definitions and application. *The Quarterly Review of Biology*, 81(3), 237–258. https://doi.org/10.1086/506237
Jelinski, D. E. (2005). There is no mother nature—There is no balance of nature: Culture, ecology and conservation. *Human Ecology*, 33(2), 271–288. https://doi.org/10.1007/s10745-005-2435-7
Jelinski, D. E. (2010). On the notions of mother nature and the balance of nature and their implications for conservation. In D. G. Bates & J. Tucker (Eds.), *Human Ecology: Contemporary Research and Practice* (pp. 37–50). Boston, MA: Springer US.
Koch, A. M. (1993). Rationality, romanticism and the individual: Max Weber's "modernism" and the confrontation with "modernity". *Canadian Journal of Political Science / Revue canadienne de science politique*, 26(1), 123–144.

Kricher, J. (2009). *The balance of nature: Ecology's enduring myth*. Princeton, NJ: Princeton University Press.

Ladle, R. J., & Gillson, L. (2009). The (im)balance of nature: A public perception time-lag? *Public Understanding of Science, 18*(2), 229–242.

Latour, B. (1996). On actor-network theory: A few clarifications. *Soziale Welt, 47*, 369–381.

Latour, B. (2005). *Reassembling the social: An introduction to actor-network theory*. Oxford, UK: Oxford University Press.

Latour, B. (2012). *We have never been modern*. Cambridge, MA: Harvard University Press.

Law, J. (2009). Actor network theory and material semiotics. In *The new Blackwell companion to social theory* (pp. 141–158). Oxford, UK: Wiley-Blackwell.

Lewontin, R., & Levins, R. (2007). *Biology under the influence: Dialectical essays on the coevolution of nature and society*. New York, NY: NYU Press.

Manuel-Navarrete, D., & Buzinde, C. N. (2010). Socio-ecological agency: From 'human exceptionalism' to coping with 'exceptional' global environmental change. In *The international handbook of environmental sociology* (pp. 136–149). Northampton, MA: Edgar Elgar Publishing.

Merchant, C. (2013). *Reinventing Eden: The fate of nature in western culture*. New York, NY: Routledge.

Mol, A. (1999). Ontological politics. A word and some questions. *The Sociological Review, 47*(S1), 74–89. https://doi.org/10.1111/j.1467-954X.1999.tb03483.x

Morton, T. (2007). *Ecology without nature: Rethinking environmental aesthetics*. Cambridge, MA: Harvard University Press.

National Research Council. (1996). *National Science Education Standards*. Washington, DC: The National Academies Press.

Newport, F. (2014). *In U.S., 42% believe creationist views of human origins*. Retrieved from http://www.gallup.com/poll/170822/believe-creationist-view-human-origins.aspx

Next Generation Science Standards Lead States. (2013). *Next Generation Science Standards: For states, by states*. Washington, DC: The National Academies Press.

Nhanenge, J. (2011). *Ecofeminism: Towards integrating the concerns of women, poor people, and nature into development*. Lanham, MD: University Press of America.

Pickett, S. T. A. (2013). The flux of nature: Changing worldviews and inclusive concepts. In R. Rozzi, S. T. A. Pickett, C. Palmer, J. J. Armesto, & J. B. Callicott (Eds.), *Linking ecology and ethics for a changing world: Values, philosophy, and action* (pp. 265–279). Dordrecht, The Netherlands: Springer.

Sagoff, M. (2003). The plaza and the pendulum: Two concepts of ecological science. *Biology and Philosophy, 18*(4), 529–552. https://doi.org/10.1023/a:1025566804906

Schizas, D. (2012). Systems ecology reloaded: A critical assessment focusing on the relations between science and ideology. In G. P. Stamou (Ed.), *Populations, biocommunities, ecosystems: A review of controversies in ecological thinking* (Vol. 101, pp. 67–92). Oak Park, IL: Bentham Science Publishers.

Schmitz, O. J. (2016). *The new ecology: Rethinking a science for the Anthropocene*. Princeton, NJ: Princeton University Press.

Scoones, I. (1999). New ecology and the social sciences: What prospects for a fruitful engagement? *Annual Review of Anthropology, 28*, 479–507. https://doi.org/10.2307/223403

Sharma, A. (2012). Global climate change: What has science education got to do with it? *Science & Education, 21*(1), 33–53. https://doi.org/10.1007/s11191-011-9372-1

Sharma, A., & Anderson, C. (2009). Recontextualization of science from lab to school: Implications for science literacy. *Science & Education, 18*(9), 1253–1275.

Simberloff, D. (2014). The "balance of nature" – evolution of a Panchreston. *PLoS Biology*. https://doi.org/10.1371/journal.pbio.1001963

Smith, N. (1984). *Uneven development: Nature, capital, and the production of space*. Oxford, UK: Blackwell.

Steffen, W., Crutzen, P. J., & McNeill, J. R. (2007). The Anthropocene: Are humans now overwhelming the great forces of nature. *AMBIO: A Journal of the Human Environment, 36*(8), 614–621.

Steffen, W., Grinevald, J., Crutzen, P., & McNeill, J. (2011). The Anthropocene: Conceptual and historical perspectives. *Philosophical Transactions of the Royal Society A: Mathematical, Physical and Engineering Sciences, 369*(1938), 842–867. https://doi.org/10.1098/rsta.2010.0327

Stone-Jovicich, S. (2015). Probing the interfaces between the social sciences and social-ecological resilience: Insights from integrative and hybrid perspectives in the social sciences. *Ecology and Society, 20*(2). https://doi.org/10.5751/ES-07347-200225

Swyngedouw, E. (2015). Depoliticized environments and the promises of the Anthropocene. In R. Bryant (Ed.), *The international handbook of political ecology* (p. 131). Cheltenham, UK: Edward Elgar Publishing.

Williams, R. (1985). *Keywords: A vocabulary of culture and society*. Oxford, UK: Oxford University Press.

Wu, J. (2011). Integrating nature and culture in landscape ecology. In S.-K. Hong, J.-E. Kim, J. Wu, & N. Nakagoshi (Eds.), *Landscape ecology in Asian cultures* (pp. 301–321). Tokyo, Italy: Springer Japan.

Zimmerer, K. S. (1994). Human geography and the "new ecology": The prospect and promise of integration. *Annals of the Association of American Geographers, 84*(1), 108–125.

CHAPTER 3

The Intended Curriculum: Locating Nature in the Science Standards

As things stand, thinking and working with education standards have become an unalienable part and parcel of the daily work of most teachers in the United States. These written prescriptions about student learning and academic performance guide their instruction and provide the foundation for the assessment of their work and of student academic progress. Though it may appear that the use of education standards is a recent invention, the fact is that they "have been expressed through laws, common curriculum and textbooks, and entrance requirements for more than 200 years" (Goertz, 2010; p. 53). Of course, their form, type, target, and use changed with time. Thus, for instance, while textbooks determined what students were expected to learn in each grade in the nineteenth century, and "education boards" set the education standards in the first half of the twentieth century (Ravitch, 2010), these days standards enter classrooms in the form of government-mandated official "standards documents" from the state departments of education or a national nongovernment organization, such as Achieve, Inc. or the National Research Council.

Until recently, "the legacy of US education embedded within our federalist construct allowed individual schools, teachers, and textbook publishers to dictate what is taught in schools" (Wixson, Dutro, & Athan, 2003; p. 70). In the last three decades, however, education standards have emerged as the lynchpin in the education reform efforts and have come to acquire a more coercive power in influencing teachers' work. In fact, efforts to raise academic standards in schools have often been labeled as "standards-based reforms" as they center on establishing clear goals for student achievement "through the establishment of standards and related assessments, generate data to improve teaching and learning, create incentives for change through rewards and sanctions, and provide assistance to low-performing schools" (Goertz, 2009, p. 206). Thus, in science, as in other content areas, an exploration of academic standards is critical to any effort to understand what counts as "official" content knowledge in K-12 settings. Academic standards embody the written formal curricula sanctioned by state and local school boards, and as such, these documents represent "what older generations choose to tell younger generations" as they struggle to define themselves and the world (Pinar, 2012). This act of "telling" is complicated, controversial, and imbued with serious ramifications for schools' role in shaping children's understanding of the world. As this chapter will show, this authoritative monologue is also deeply political as it prioritizes and legitimizes a selective ideological interpretation of the world that serves the interests of a few over the rest.

In this chapter, we explore the current iterations of science content standards to delineate the broad contours of the officially sanctioned representation of the natural world in school science knowledge. According to Wixson, Dutro, and Athan (2003), "The story of content standards is a national story, a state story, a story of specific disciplines, and a story of philosophical and theoretical shifts and differences that have had an impact on views of teaching and learning across disciplines" (p.69). Our analysis highlights the different strands of this story through a critical examination of the national *Next Generation Science Standards* (National Research Council, 2013) as well as the current (*Georgia Performance Standards*) and upcoming (*Georgia Standards for Excellence*) science content standards in the state of Georgia (Georgia Department of Education, n.d., n.d.). We use the example of Georgia both because it is the context in which we are currently working and because Georgia seems typical of the majority of states that have "adapted" rather than "adopted" the NGSS in a balancing act between following national trends and maintaining state-level autonomy over education. We begin with a brief description of our methodology and the methods we adopted to this end.

Critical Discourse Analysis

We are talking here about textual representations of the natural world in the science content standards. Therefore, we adopted the methodological framework of critical discourse analysis (CDA) as it assumes that "every practice has a semiotic element" (Fairclough, 2004; p. 122), and a critical examination of these elements can uncover the ways in which powerful authoritative texts work to privilege and hegemonize some representations of the world over others. In CDA the focus is on those semiotic elements of social practice that are marked by distinctly identifiable patterns of language in use, in speech as well as writing (Fairclough, 2003; Fairclough & Wodak, 2004). These identifiable patterns of language in use when seen as elements of social life are labeled as *discourses*.

According to Mazid (2014), "there is no single theory or method which is uniform and consistent throughout CDA studies" (p. 18). However, over time, different CDA frameworks have emerged that focus on different aspects of discourses and their relationship with social life. In this chapter and the chapters to follow, we follow Fairclough's textually oriented approach to critical discourse analysis (TODA) that follows a three-dimensional framework for analyzing discourse "where the aim is to map three separate forms of analysis onto one another: analysis of (spoken or written) language texts, analysis of discourse practice (processes of text production, distribution and consumption) and analysis of discursive events as instances of socio-cultural practice" (Fairclough, 1995; p. 2).

As an element of social life, discourse makes its presence felt by shaping (a) the genres (the ways people linguistically act and interact in communication with others), (b) discourses (the ways the social and material world is represented in speech as well as writing), and (c) styles (the ways the social and personal identities are linguistically constituted) in communication (Fairclough, 2003). Because of our focus on representations of the natural world, our analysis centers on teasing out the different discourses in science content standards. However, we do make a brief comment on the genre of writing typical of science content standards to highlight its dis-embedded and abstract nature that corresponds very well with the key theme of naturalization of neoliberal rationality that emerged from the discursive analysis of the standards documents.

For identification of the different discourses in these texts, we adopted a twofold strategy. First, because value systems and assumptions about the world are often closely associated with particular discourses, we began with an analysis of assumptions in the content standards. We felt this was important to do because implicit and explicit assumptions in oral or written texts constitute the "common ground" that is assumed to be shared among the interlocutors and by which any social communication or interaction gets done. We, therefore, looked for four main types of assumptions in the standards documents: (i) *existential assumptions* about what exists; (ii) *propositional assumptions* about what is, can be, or will be the case; (iii) *logical implications* that "can be logically inferred from the features of language" (p. 60); and (iv) *value assumptions* about what is good or desirable (Fairclough, 2003).[1] Next, through an iterative process of repeated reading and textual analysis we identified "the main parts of the world (including areas of social life) which are represented – the main themes" and "the particular perspective or angle or point of view from which they are represented" (Fairclough, 2003; 129). We assume that what is written is dictated by how it gets written and who is involved in the act of writing. Therefore, before we undertake a discursive dive into the science content standards documents, it might be a good idea to first get an understanding of how they got written.

SCIENCE CONTENT STANDARDS: A BRIEF STORY OF THEIR ORIGINS

The current wave of standards-based reforms in the United States began in the 1990s with the stated purpose of raising academic standards and moving public education from a putative low-level education, focused on basic skills and behavioristic thinking, to one that was guided by a more academically rigorous "thinking curriculum" based on sociocognitive views of teaching and learning (Wixson, Dutro, & Athan, 2003). These reforms were given a broad political legitimacy by the political and economic elite who were able to persuade the public that the nation was at risk of losing its economic power and prosperity because of the supposedly poor quality of education offered by its public schools (Berliner & Biddle, 1995). In fact, barring a few dissenting voices, these reforms were supported by influential forces from all sides of the political spectrum (Shepard, 2015). Helped by such a supportive climate, the National

Research Council was able to bring together science teachers, scientists, administrators, teacher educators, school board members, parents, and others to work on national science standards that could shape science education in the United States with a vision of making "scientific literacy for all a reality in the 21st century" (Committee on Development of an Addendum to the National Science Education Standards on Scientific Inquiry, 2000). After a sustained collaboration of about 3 years, the *National Science Education Standards (NGSS)* were unveiled to the public in 1995. These standards played a significant role in influencing the intended science curricula throughout the United States as they were used as the template by most states to write their own state-level science standards at that time (National Research Council, 2001). Of course, the efforts at the state level were also partly driven by the No Child Left Behind Act of 2001, which tied federal funds for K-12 education programs to new standards, such that "state education agencies were required to develop and implement standards and align assessments to monitor student learning" (Tran, Reys, Teuscher, Dingman, & Kasmer, 2016; p. 120).

Between the mid-1990s and the mid-2000s, all states developed their own science content standards, with the National Science Education Standards remaining the leading visionary document for these efforts at both the national and state level. But by the turn of the decade in 2010, the social efficiency driven narrative of declining quality of public education, and the threat that it posed for the nation's economic and technological superiority had again been resurrected and brought to the forefront of policymaking in education by the political and economic elite (Sharma, 2016). A privately funded effort shepherded by the Carnegie Corporation of New York laid the groundwork for another round of revisions to the science content standards for the nation's schools. As usually happens when reforms are initiated, the first step was taken by creating a strong argument for reform. Carnegie Corporation did this in 2007 by commissioning a report *The Opportunity Equation: Transforming Mathematics and Science Education for Citizenship and the Global Economy*. This report was authored by a distinguished panel of scientists, academicians, and public and private leaders. Though the report gave a rhetorical nod to the importance of quality science and math education for social mobility and the nation's democracy, its key focus remained on improving "the nation's capacity to innovate for economic growth and the ability of American workers to thrive in the global economy" by placing "math and science more squarely at the center of the educational enterprise" (Commission on Mathematics and Science Education, 2009; p. vii).

With this report as the basis, the Carnegie Corporation then supported the National Research Council (NRC) in partnership with the American Association for the Advancement of Science (AAAS), National Science Teachers Association (NSTA), and Achieve, the primary writers of the Common Core State Standards for Language Arts and Mathematics, to develop *A Framework for K-12 Science Education* in 2011 (National Research Council, 2012; Pruitt, 2014). This framework articulated a vision "of the scope and nature of the education in science, engineering, and technology needed for the 21st century" and outlined a set of practices, cross-cutting concepts, and disciplinary core ideas that integrated engineering and technology with science so as "to reflect the importance of understanding the human-built world and to recognize the value of better integrating the teaching and learning of science, engineering, and technology" (National Research Council, 2012; p. 8). The framework was intended as a guide for the final step of writing "a next-generation set of science standards for voluntary adoption by states" (p. 8)—a process that was coordinated by Achieve, an independent, nonprofit education reform organization founded in 1996 by a group of state governors and business leaders. The development of these standards was a collaborative effort of representatives from state departments of education of 26 states, K-12 educators, higher education faculty, state science supervisors, scientists, engineers, researchers, and science teachers. Georgia was one of the states that participated in this process. The new science standards, labeled as *Next Generation Science Standards*, were finally released in 2013 (National Research Council, 2013). As of 2017, 19 states have adopted the NGSS, while an additional 24 states have adopted new science standards that are adaptations of the NGSS.

It has been noted by many scholars that in recent decades the global economic and political elite has made a sustained effort to reshape public education such that it is primarily designed to serve an overall neoliberal agenda (Macrine, McLaren, & Hill, 2010). A reflection of such efforts can be seen in the stated rationale of the Next Generation Science Standards. The official website "Next Generation Science Standards: For States, by States" for these standards highlights the following four main reasons that these standards are needed: (a) reduction of the United States' competitive economic edge, (b) lagging achievement of US students,

(c) the importance of science education for all careers in the modern workforce, and (d) the need of scientific and technological literacy for an educated society ("The Need for Standards"; n.d.). The *Framework for K-12 Science Education* that guided the development of these standards while establishing the rationale for science education reform acknowledged that in addition to the imperative of keeping "the United States competitive in the international arena," a compelling case can also be made that "understanding science and engineering, now more than ever, is essential for every American citizen." However, as is evident from the aforementioned officially stated reasons for the Next Generation Science Standards, the social efficiency imperative that views schools as little more than training camps for future workers was clearly the dominant influence in shaping the new standards. Our analysis of the standards that follows also supports this thesis.

As mentioned earlier, the state department of education of Georgia took part in the development of the new science standards. However, when the time came to adopt these standards, the state chose to adapt them instead, so as to be able to claim that the state was developing their own "Georgia-owned and Georgia grown" science standards, called the *Georgia Standards for Excellence (GSE)* (Georgia Department of Education, n.d.). Though the *Georgia Standards for Excellence* are based on *A Framework for K-12 Science Education* and follow a similar format to the NGSS by integrating practices and concepts into each standard, developers claim that the GSEs also draw inspiration from Project 2061's *Benchmarks for Science Literacy* as well as Georgia-specific themes and resources ("Georgia Department of Education: Science", n.d).

These standards went into effect in all Georgia public schools during the academic year 2017–18. They replace the *Georgia Performance Standards (GPS)* that have served as the intended curriculum for Georgia schools since 2004[2] ("Georgia Department of Education: Georgia Performance Standards", n.d.). According to the Georgia Science Teachers Association which played a leading role in the development of the *Georgia Standards for Excellence*, the new science standards address two main limitations of the previous *Georgia Performance Standards*. First, they fully integrate the *Characteristics of Science* standards that include the *Habits of Mind* and *Nature of Science* standards that had been separated from the science content standards in the prior standards, so that teachers are not "pressured into sacrificing deep, contextualized learning through investigations in favor of superficial coverage and memorization of content." The need to integrate practices and characteristics of science with science content was, in fact, the most significant feedback from Georgia's science teachers in the preliminary teacher survey conducted prior to the revision process (https://www.georgiascienceteacher.org/Next-Gen-Updates/3498157). Second, the science content was revised to bring it up to date with current scientific knowledge and the needs of the students, society, and economy, because as the Georgia Science Teachers Association affirms, "The issues we face as a society, the economy for which we are preparing our students, and our understanding of how students best learn science have all changed dramatically in the last 20 years" (Georgia Science Teachers Association, n.d.). In addition, the new standards include elements that relate to engineering practices, a major emphasis of the *Framework for K-12 Science Education*.

Both the *Next Generation Science Standards* and *Georgia Standards for Excellence* were an outcome of a sustained collaborative process that included many stakeholders, such as science teachers, post-secondary educators, business and industry representatives, parents, and state educational agencies and nongovernment organizations. The actual writing of the standards in both cases was primarily performed by the K-12 educators and higher education faculty. The members of the writing teams were selected on the basis of recommendations from groups like National Science Teachers Association (in case of NGSS) and Georgia Science Teachers Association (in case of GSE). All of these members have impressive resumes, and their expertise in matters of science education can scarcely be doubted. However, it is also interesting to note that some key scientific disciplines are poorly represented in these working groups, for example, ecologists and environmental scientists are conspicuously absent from the writing teams for both sets of standards ("Writing Team", n.d.). It must be conceded, though, that the writing team for the *Framework for Science Education* that guided the writing of both standards documents did include the well-known ecologist Rodolfo Dirzo (National Research Council, 2012). The precise nature of the relationship is a matter of conjecture, but it can't be denied that the composition of the writing team and the process of development of these standards would have a strong bearing on how the world is represented in these curriculum documents. We will revisit this issue towards the end of the chapter. But first let us understand how the natural world is represented in science content standards.

The World as Seen Through Science Standards

As mentioned, we conducted a critical discourse analysis of three sets of science content standards: *Next Generation Science Standards (NGSS)* at the national level and *Georgia Performance Standards (GPS)* and *Georgia Standards for Excellence (GSE)* at the state level. Before we go into details, here is a broad overview of these curricular documents.

Next Generation Science Standards (NGSS) are student performance expectations written in the form of assessable statements of what students should be able to do in order to demonstrate that they have met the standards. These standards are for all students and it is expected that by the time students leave high school all of them would be able to achieve proficiency with respect to all the performance expectations in the NGSS ("How to Read the Next Generation Science Standards", 2013). Each of these performance expectations incorporates three dimensions pulled from the *Framework on Science Education*:

(a) A core disciplinary idea: There are a total of 44 core disciplinary ideas derived from four disciplinary areas of physical sciences, life sciences, Earth and space sciences, and engineering, technology, and applications of science. These core ideas drawn from within and across the disciplines were chosen "to prepare students with sufficient core knowledge so that they can later acquire additional information on their own" ("The Next Generation Science Standards: Executive Summary", 2013) and also to "avoid shallow coverage of a large number of topics" so that students are able to explore each idea in greater depth (National Research Council, 2012; p. 11). An example of core disciplinary idea would be ESS3.D—Global Climate Change: *How do people model and predict the effects of human activities on Earth's climate?* In the framework document this question encapsulating the core disciplinary idea of global climate change is elaborated upon in a few paragraphs.

(b) A science or engineering practice: Guided by the ideas advocated in the framework document that science education should expose students to science as practiced in the real world, and technology and engineering should be integrated into the science standards, NGSS includes eight science and engineering practices, such as *Analyzing and Interpreting Data* (SEP4) that are woven through the student performance expectations.

(c) A cross-cutting concept: These are concepts that "unify the study of science and engineering through their common application across fields" (National Research Council, 2012; p. 2). The standards incorporate seven such concepts, for instance, *cause and effect (CC2)*.

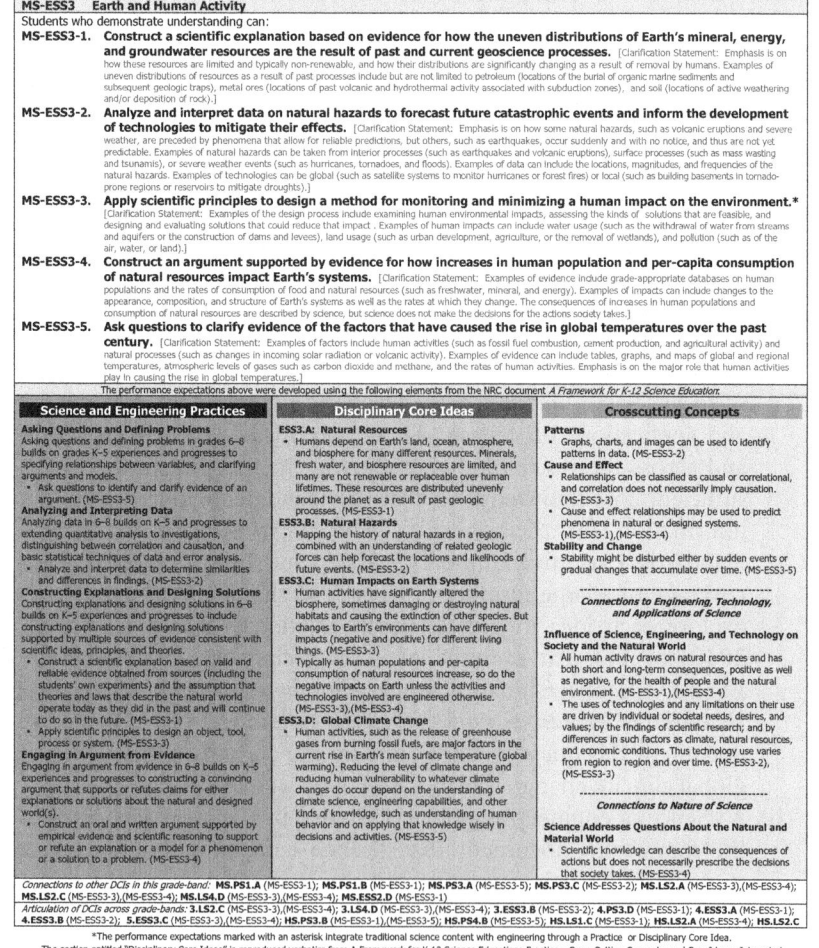

Fig. 3.1 Next Generation Science Standards

All NGSS content standards begin with a declaration that "Students who demonstrate understanding can: ...." This is then followed by expected performance, such as "construct an explanation," "create ...," and "evaluate ...." For instance, for the disciplinary core idea—*MS-LS2 Ecosystems: Interactions, Energy, and Dynamics*—we find five performance standards. One of which is "Students who demonstrate understanding can: Evaluate competing design solutions for maintaining biodiversity and ecosystem services" (MS-LS2-5). Many of the performance standards are then followed by a clarification statement to further clarify the intent and focus of the concerned performance standard (see Fig. 3.1).

Coming to content standards for the state of Georgia, we find that currently the situation is in a state of transition. 2016–17 was the last year for the previous science standards, the *Georgia Performance Standards* (GPS). From 2017–18 onwards, Georgia began implementing the new *Georgia Standards of Excellence* (GSE). From kindergarten through eighth grade, both GPS and GSE cover disciplinary areas of life science, Earth and space science, and physical science. For high school, one finds GPS and GSE standards for biology, chemistry, Earth systems, environmental science, physical science, and physics courses (Fig. 3.2).

Based on Project 2061's *Benchmarks for Science Literacy* and aligned with the National Research Council's *National Science Education Standards*, the Georgia Performance Standards (GPS) takes the stand that to achieve scientific literacy, students need to understand the "Characteristics of Science" as well as science "Content." The instruction should be organized so that both are integrated in lesson plans and teaching practices. Thus, in the standards documents for different grades (K-8) or courses (High School), "Characteristics of Science" and science "Content" are presented as co-requisites. In addition, GPS for each grade or course also include a reading standard to enhance reading across disciplinary boundaries. The "Characteristics of Science" standards include "Habits of Mind" standards, such as "Students will communicate scientific ideas and activities clearly" (S7CS6), and "Nature of Science" standards, such as "Students will investigate the features of the process of scientific inquiry" (S7CS9). Unlike the NGSS, however, in GPS, the "Characteristics of Science" standards are presented separately from the content standards, rather than as integrated standards. The result of this has been that while the Characteristics of Science standards are presented as required to be taught concurrently with the content standards, in actual classroom practice, many teachers have viewed these as optional standards that are rarely

Co-Requisite - Content

S7L1. Students will investigate the diversity of living organisms and how they can be compared scientifically.
  a. Demonstrate the process the development a dichotomous key.
  b. Classify organisms based on physical characteristics using a dichotomous key of the six kingdom system (archaebacteria, eubacteria, protists, fungi, plants, and animals).

S7L2. Students will describe the structure and function of cells, tissues, organs, and organ systems.
  a. Explain that cells take in nutrients in order to grow and divide and to make needed materials.
  b. Relate cell structures (cell membrane, nucleus, cytoplasm, chloroplasts, mitochondria) to basic cell functions.
  c. Explain that cells are organized into tissues, tissues into organs, organs into systems, and systems into organisms.
  d. Explain that tissues, organs, and organ systems serve the needs cells have for oxygen, food, and waste removal.
  e. Explain the purpose of the major organ systems in the human body(i.e., digestion, respiration,reproduction, circulation, excretion, movement, control, and coordination, and for protection from disease).

S7L3. Students will recognize how biological traits are passed on to succesive generations.
  a. Explain the role of genes and chromosomes in the proess of inheriting a spcific trait.
  b. Compare and contrast that organisms reproduce asexually and sexually (bacteria, protists, fungi, plants & animals).
  c. Recognize that selective breeding can produce plants or animals with desired traits.

S7L4. Students will examine the dependence of organisms on one another and their environments.
  a. Demonstrate in a food web that matter is transferred from on organism to another and can recycle between organisms and their environments.
  b. Explain in a food web that sunlight is the source of energy and that this energy moves from organism to organism.
  c. Recognize that changes in environmental conditions can affect the survival of both individuals and entire species.
  d. Categorize relationships between organisms that are competitive or mutually beneficial.
  e. Describe the characteristics of Earth's major terrestrial biomes (i.e. tropical rain forest, savannah, temperate, desert, taiga, tundra, and mountain) and aquatic communities (i.e. freshwater, estuaries, and marine).

S7L5. Students will examine the evolution of living organisms through inherited characteristics that promote survival of organisms and the survival of successive generations of their offspring.
  a. Explain that physical characteristics of organisms have changed over successive generations (e.g. Darwin's finches and peppered moths of Manchester)
Georgia Department of Education
Kathy Cox, State Superintendent of Schools
8/29/2006 2:52PM Page 7 of 8
All Rights Reserved

Fig. 3.2   Georgia Performance Standards

assessed and can therefore be safely ignored. Considered linguistically, the verbs used for describing the expected performances in the GPS do not highlight science practices in the same way as in NGSS. For instance, a content standard for the high school Environmental Science course is written as: "Students will *recognize* (emphasis ours) that human beings are part of the global ecosystem and will evaluate the effects of human activities and technology on ecosystems" (SEV5) (refer to Appendix A: Referenced Science standard statements). In the six sub-elements that accompany this standard, the performance descriptors—"describe" and "explain—fail to guide teachers and students to take up the characteristics of science.

Unlike the GPS, the new Georgia Standards of Excellence (GSE) are more precise in the description of expected student performance after instruction. First, all GSE standards begin with "Obtain, evaluate, and communicate information to" as a lead-in to the expected performance. Second, the performance descriptors in the sub-elements accompanying each standard are all related to science and engineering practices, similar to the approach taken in NGSS. For instance, performance standard SES6 for sixth-grade science is given as: "Obtain, evaluate, and communicate information about how life on Earth responds to and shapes Earth's systems" (refer to Appendix A: Referenced Science standard statements). Third, this is followed by detailed elaborations of expected performance in the accompanying sub-elements, such as "Construct an argument from evidence that describes how life has responded to major events in Earth's history (e.g., major climatic change, tectonic events) through extinction, migration, and/or adaptation" (refer to Appendix A: Referenced Science standard statements). Like NGSS, some of the sub-elements of performance standards are accompanied by a clarification statement to help teachers deconstruct the intent and focus of the concerned standard statement. Other important changes in GSE are (a) while "characteristics of science" and "nature of science" were given separately from content standards in GPS, they have now been embedded in the content standards in the GSE. (b) Standards related to reading in content areas have been removed in GSE. (c) An effort was made to make the science content more rigorous and up to date with current scientific understanding in respective disciplines (Fig. 3.3).

A curriculum may indeed be an extraordinarily complicated conversation, as Pinar (2012) opined. But the way this conversation is carried out between those who write and those who use academic standards has by

SEV2. Obtain, evaluate, and communicate information to construct explanations of stability and change Earth's ecosystems.
  a. Analyze and interpret data related to short-term and long-term natural cyclic fluctuations associated with climate change.
      (*Clarification statement:* Short-term examples include but are not limited to Niño and volcanism. Long-term examples include but are not limited to variations in Earth's orbit such as Milankovith cycles.)
  b. Analyze and interpret data to determine how changes in atmospheric chemistry (carbon dioxide and methane) impact the greenhouse effect.
  c. Construct an argument to predict changes in biomass, biodiversity, and complexity within ecosystems, in terms of ecological succession.
  d. Construct an argument to support a claim about the value of biodiversity ecosystem resilience including keystone, invasive, native, endemic, indicator, and endangered species.

Fig. 3.3 Georgia Standards of Excellence

now become stable and similar across states. It is important to understand the nature of this conversation as that will shed light on how certain representations of our world come to acquire their hegemonic stature in the science classrooms. An analysis of the genre of the text in NGSS, GSE, and GPS standards documents proved very helpful in this regard. This is because, a genre is "a socially ratified way of using language in connection with a particular type of social activity" (Fairclough, 1995; p. 14). Thus, understanding the genre of science standards showed to us how these standards contribute to the production of teaching and learning in science classrooms. We specifically looked at the level of abstraction, the purpose as revealed by the generic structure of the texts, communication technology, and the implied social relations in these texts.

If you peruse these documents, you will find that despite some content-related differences, they read quite similarly. First, they are all written in a very abstract language. From a typological stance, these documents would fit within the science genre of whole-parts compositional report as they deal with content organized by classification and composition (Martin & Rose, 2008). In these reports, we find science content organized and classified taxonomically according to grade level, disciplinary areas, and disciplinary core ideas. Further, these documents present un-localized, spatiotemporally independent content in an unchanging format. In the Georgia standards, despite the rhetoric of "Georgia-owned and Georgia grown," there is little content that is specific to the state of Georgia. Likewise, there is nothing in the NGSS that would be out of place in a science classroom anywhere in the world. Further, when we look at the content, we find that it is accompanied by instructions for learning and teaching that are defined by context-independent and abstract special expressions, such as "evaluate claims, evidence and reasoning," "investigation," "analyze and interpret data," "construct an argument," and so on, that only fluent practitioners of this genre would find familiar.

Technical taxonomies and special expressions are, of course, only two of the many features of the genre that contribute to making scientific language abstract and difficult to comprehend. The abstract nature of the text is also emphasized by the high lexical density achieved by packing a lot of content through nominalization in each standard statement. For instance, consider the following GSE environmental science standard SEV1: "Obtain, evaluate, and communicate information to investigate the flow of energy and cycling of matter within an ecosystem" (refer to Appendix A: Referenced Science standard statements). To understand this standard, a teacher would have to unpack the embedded meaning of processes nominalized as verbs, such as "obtain" "evaluate," and "communicate," as well as a circumstance nominalized by the prepositional phrase "within an ecosystem." This would certainly be no easy task as understanding such academic standards calls for a deep familiarity with both pedagogy and science content.

These features of the genre of science standards work to dis-embed the genre from its moorings. According to Fairclough (2003), "The disembedding of genres is a part of the restructuring and rescaling of capitalism" (p. 69). We find ourselves in agreement with this assessment because it is evident that the dis-embedment of genre in these texts allows them to work as "social technology" tools for the purposes of curricular control through centralization of curricular writing and standardization of science content across geographical and temporal boundaries. This enables giant corporations like Pearson and Educational Testing Services to create a vast and highly profitable market on standardized assessments and curricular material needed for successful translation of these academic standards into enacted curriculum. If we look around, we will find many examples of similar dis-embedment in our globalized social lives. For instance, money has been dis-embedded from its physical manifestations as in bank notes, information is no longer embedded in physical books and papers, and even war is now being increasingly dis-embedded from spatiotemporal locations through drones and cyber-warfare (Eriksen, 2014). As we report ahead, we find similar dis-embedment in representation of nature in these standards. These correspondences can't be happenstance. In fact, dis-embedment can be seen as a key stage in the commodification of our world, whether the entity undergoing commodification resides in an Amazonian jungle or on the pages of a curricular document.

Curricular control through standards documents is also aided by the nature of strategic action and social relations implicit in the generic structure of the text. All the standards whether they belong to NGSS, GSE, or GPS begin with action verbs reflecting measurable learning performances that students are expected to demonstrate as a result of learning science. Examples are develop, use, compare, analyze, interpret, evaluate, plan and investigate, interpret, construct, and so on. While sometimes broad in scope, these action verbs are authoritative in nature and position teachers and students as passive nodes acted upon by distant and more agential centers of authority. The social hierarchy in this communication is clear and puts teachers and students at a great disadvantage in terms of power relations between them and the curriculum makers. Further, the originators and recipients of the communication are missing from the text. As a result, there is no attempt to lessen the social distance between standard writers and standard implementers. The one-way mediated (via text) nature of technology used in delivering the communication of these standards to its intended recipients also removes any possibility of democratic dialogue about what is present or absent from the curriculum. With this appreciation of the ways in which the genre of these standards documents structures the nature of action and interaction on curriculum, we are now in good standing to explore the ways in which nature is represented in these documents.

Assumptions About the World

We start with the assumptions that these standards documents make about the world. According to Fairclough (2003), "What is 'said' in a text is 'said' against a background of what is 'unsaid', but taken as given" (p. 40). This implicitness is an important property of all texts as it carries with it not only the ontological commitments that the text makes about the world, but also the values it tacitly communicates to the audience. Uncovering these presuppositions in the text becomes important because "presuppositions have remarkable properties regarding the triggering of audience consent to the message expressed. Presupposed content is, under ordinary circumstances, and unless there is a cautious interpretive attitude on the part of the hearer, accepted without (much) critical attention (whereas the asserted content and evident implicatures are normally subject to some level of evaluation)" (Wodak, 2007; p. 214). Thus, let us first get acquainted with the ontological foundations and values that are not so easily discerned, but on the basis of which the authoritative representations of the world are conveyed to the teachers and students in these documents. Our analysis of assumptions uncovered the ontological assumptions about "what there is" and guiding values about "what should be." We begin first with ontological assumptions.

Ontology: What There Is?
Human Impact: All the three sets of standards make assumptions which indicate an acceptance of the view that humans have significantly impacted the world. In each document the world is taken to be understood as the natural world that is devoid of humans. Further, the texts do not differentiate between different human communities or societies. All human beings are lumped together under one category as "humans" as if all individuals are equally responsible for the "human impact." For instance, GSE high school Environmental Science content standard SEV4 wants students to be prepared so that they can "Obtain, evaluate, and communicate information to analyze human impact on natural resources" (refer to Appendix A: Referenced Science standard statements). Both sets of Georgia standards give some examples of human impact for the benefit of teachers, such as smog, ozone depletion, urbanization, ocean acidification, habitat destruction, and depletion of soil fertility. Except for ozone depletion, these are perhaps all good relevant examples of human impact. However, we also couldn't help but notice the conspicuous absence of climate change as an example here.[3]

Change and Stability The issue of stability and change in Earth's systems is important in these documents. It is a common assumption across the three standards that even though ecosystems change, their fundamental processes and attributes are to be understood independent of space and time; see, for example, HS-LS2-4 (NGSS), SEC3 (GPS), and SEV1 (GSE) (refer to Appendix A: Referenced Science standard statements). We also noticed that while stability is the default status as regards Earth's systems in elementary and middle grades, the issue of change becomes more prominent in higher grades. All three sets of standards make different though compatible assumptions about change to ecosystems. In NGSS it is assumed that stability is the norm in ecosystems when conditions are stable. However, changes in conditions can cause an ecosystem to become a different system (HS-LS2-6) (refer to Appendix A: Referenced Science standard statements). Both GSE and GPS take a different tack and focus more on understanding ecosystems under stable conditions. When ecosystems change, it is assumed in both Georgia standards that they can be understood on the basis of a linear singular model of ecological succession; see, for example, SEV2(c) (GSE) and SEC3 (e) (GPS) (refer to Appendix A: Referenced Science standard statements). An important context for understanding change to Earth's systems is, of course, climate change. All three sets of standards assume that climate change is real. Further, NGSS and GSE assume that it has both natural and anthropogenic causes. Though GSE does not prioritize between these two sources of climate change, NGSS does call out humans as the major driver of rise in global temperatures. GPS, on the other hand, assumes that climate change is just one among environmental issues facing the planet, see for instance SEC5 (GPS) (refer to Appendix A: Referenced Science standard statements).

World as a System Systems theory is a powerful perspective in ecology and Earth science. It assumes that our world can be best understood as a complex system comprising interrelated and interdependent parts in which interacting biological, chemical, and physical processes combine to exhibit emergent properties. Further, systems perspective makes good use of concepts and laws from thermodynamics to explain flow of energy within and across systems. All three standards documents represent the world as a system; see, for instance, HS-ESS2-4 (NGSS), SES1 (GSE), and SEV2 (GPS) (refer to Appendix A: Referenced Science standard statements). Further, in this perspective, life in "natural systems" is assumed to be hierarchically organized into progressively higher levels of organization, such as in sixth-grade GPS standard SEV2: "Recognize and give examples of the hierarchy of the biological entities of the biosphere (organisms, populations, communities, ecosystems, and biosphere)" (refer to Appendix A: Referenced Science standard statements). In this hierarchical system causal influences only flow down from the external inputs and environmental conditions to the populations and individuals. For instance, in middle school NGSS standard MS-LS2-1, students are expected to "Analyze and interpret data to provide evidence for the effects of resource availability on organisms and populations of organisms in an ecosystem" (refer to Appendix A: Referenced Science standard statements).

The system perspective in these documents also assumes that ecosystems are the appropriate units for understanding Earth's natural systems. These ecosystems are generally assumed to be closed systems for the purposes of understanding ecological processes and phenomena. Therefore, cycling of matter and flow of energy are always presented as occurring only within ecosystems. For example, GSE high school standard SEV1 asks students to "Obtain, evaluate, and communicate information to investigate the flow of energy and cycling of matter within an ecosystem" (refer to Appendix A: Referenced Science standard statements). It should be acknowledged though that in NGSS, we do find that when it comes to systems, such as the hydrosphere or atmosphere that are at a higher level of organization than ecosystems, there is an acknowledgment of relationships including flow of energy between systems. For example, the high school NGSS standard HS-ESS2-4 expects students to "Use a model to describe how variations in the flow of energy into and out of Earth's systems result in changes in climate" (refer to Appendix A: Referenced Science standard statements).

Finally, the systems perspective in these documents also assumes that our world is best understood if it is seen as divided into two ontologically distinct realms: a "natural world" devoid of humans and a "social world" comprising only of humans. For instance, NGSS enjoins students to "Design, evaluate, and refine a solution for reducing the impacts of human activities on the environment and biodiversity" (HS-LS2-7) and "Evaluate or refine a technological solution that reduces impacts of human activities on natural systems" (HS-ESS3-4) (refer to Appendix A: Referenced Science standard statements). This standard marks a clear ontological distinction between a natural world where humans don't exist and the world of humans. Similarly, GSE positions humans in a different ontological category from the rest of life on Earth when it asks students to "Analyze and interpret data to provide evidence for how ... human activity affect individual organisms, populations, communities, and ecosystems" (S7 L4) and "Obtain, evaluate, and communicate information about the effects of human population growth on global ecosystems" (SEV5) (refer to Appendix A: Referenced Science standard statements).

Values: What Should Be?
Science texts often give the impression of being value neutral. These standards documents bearing the responsibility of conveying what children should learn as science are no different. Emphasizing its objective stance, NGSS claims that "Science knowledge indicates what can happen in natural systems not what should happen" (HS-ESS3-2; Clarification Statement) (refer to Appendix A: Referenced Science standard statements). However, on closer reading the standards documents revealed themselves to be no less value laden than any other text. Perhaps, because all three texts derive their academic wherewithal from similar and compatible resources, such as *A Framework for Science Education*, and are written by people subscribing to the mainstream consensus view of science, there is a close correspondence between these documents in terms of their underlying values. As we examined the documents, we found these values clustering around two categories as presented below.

(a) *How to value human action?* All three documents take an overall negative view of the human relationship with the natural world. Though human beings are credited with the ability to come up with technological solutions to mitigate their impact on the world, human impact is uniformly assumed as damaging to other life forms, natural resources, and Earth's systems. For instance, GSE high school standard SEV4 asks students to "Obtain, evaluate, and communicate information to analyze human impact on natural resources" (refer to Appendix A: Referenced Science standard statements). It is not obvious from the language of the standard if the human impact on natural resources is positive or negative. However, the accompanying sub-element (b) that asks students to "Design, evaluate, and refine solutions to reduce human impact on the environment" makes it evident that this relationship is assumed to be anything but positive (refer to Appendix A: Referenced Science standard statements). This negative view of human impact is par for the course as it has become a foundational belief both within mainstream science and public opinion (Anderson, 2017). What we found more interesting in the standards was a total occlusion of the impact of human activity on other humans. Human activity in relation to nonhuman parts of the world, such as mining and timber logging, often leads to serious consequences for communities that live in those parts of the world. But standards do not make any mention of it. This omission carries with the logical implication that human activity either has no effect on the lives of the people or this impact is not worthy of inclusion in the study of science. Either way, this view privileges a perspective that sees a world devoid of humans as the primary object of scientific inquiry and knowledge and assumes that the impact of human activity on fellow humans is not worthy of acknowledgment in the science curricula. As we document in the chapters ahead, such notions were common both in the classroom discourse and students' conceptions about the world.

Further, while the negative impact of humans on the rest of the world are acknowledged, all three sets of standards exhibit a positive belief in the ability of humans to come up with solutions to environmental problems and expect students to learn to take environmental action to reduce their impact on the natural world. Such actions usually have both technological and social components. However, in these standards documents we see a complete marginalization and hence devaluation of the social aspects of environmental action. That is, we see little recognition in the standards that undertaking successful environmental action is not like solving a technological puzzle on a lab bench as it also invariably entails dealing with the social aspects of environmental problems and attending to the social consequences of the intended solutions. We see a devaluation of the social in other aspects of environmental action as envisaged in the standards as well. Such actions cover a broad spectrum of available options and can range from being individual oriented to more collective in nature. For example, in all the three sets of standards, we find a privileging of individual oriented action over collective solutions. For instance, GSE (standard SEV5) expects students to learn to "Design and defend a sustainability plan to reduce your individual contribution to environmental impacts, taking into account how market forces and societal demands (including political, legal, social, and economic) influence personal choices" (refer to Appendix A: Referenced Science standard statements). In contrast, we did not come across a single standard that even vaguely alluded to the possibility or desirability of collective or democratic solutions to environmental problems.

(b) *How to value human reason?* Science standards portray school science as not just a preparation for understanding of the world as scientists do but also to solve problems like them. With the highlighting of engineering ideas and practices in both NGSS and GSE, we also see a strong preference for a technocentric-based value system in these two sets of standards, though technocentric values aren't absent from GPS either. First, all issues, even those with an integral social component, are reduced to technological issues. For instance, the management of natural resources is an intensely sociopolitical issue as decades of research in natural resource management have shown; see, for example, Nelson's (2012) edited book on the politics of natural resource management in Africa. But the high school NGSS standard HS-ESS3-3 valorizes technocentric values by reducing the issue of management of natural resources to a purely technological matter by asking students to "Create a computational simulation to illustrate the relationships among management of natural resources, the sustainability of human populations, and biodiversity" (refer to Appendix A: Referenced Science standard statements). As you can see, all sociopolitical factors have been eschewed in favor of technical considerations. Second, in accompaniment with the reduction of complex issues to technological problems, there is a clear positive valuation of the role of science and engineering in solving environmental issues. For instance, in the disciplinary core idea, ESS3.C, related to "Human Impacts on Earth's Systems" in the high school NGSS standards, it is affirmed that "Scientists and engineers can make major contributions by developing technologies that produce less pollution and waste and that preclude ecosystem degradation" (National Research Council, 2012; p. 195). Similarly, GPS high school Environmental Science standard SEV4 want students to be able to "Describe how technology is increasing the efficiency of utilization and accessibility of resources" (refer to Appendix A: Referenced Science standard statements).

Third, we also notice a tendency in these standards to fall back on an instrumental rationality for solving such problems. This gets manifested in several ways, such as in focusing on understanding *process* over *cause* and *action* over *actors* that leads to valuing "how" questions over "why" questions for solving problems. This differential valuation is clearly evident in the context of environmental issues. The issue of climate change serves as a case in point. In all three sets of standards, the focus is on understanding the processes that are involved in climate change, such as how an increase in greenhouse gases causes climate change and how that impacts the natural world, rather than why there is an increase in greenhouse gases and who are the human actors that are primarily responsible for it. For instance, consider the following sub-element in the high school Earth systems standard SES5 in GSE: "Construct an argument relating changes in global CLIMATE to variation to Earth/sun relationships and atmospheric composition" (refer to Appendix A: Referenced Science standard statements). Similarly, in the NGSS students are expected to "Use a model to describe how variations in the flow of energy into and out of Earth's systems result in changes in climate" (HS-ESS2-4) (refer to Appendix A: Referenced Science standard statements). Our conclusions, therefore, support Hufnagel, Kelly, and Henderson (2017) who, in their investigation of how the environment and environmental issues are conceptualized and positioned in the NGSS, found that agency in these standards "is more often ascribed to actions or activities rather than people and when solutions to environmental issues are included, the focus is on technoscientific solutions" (p. 1).

Along with technocentrism, we also found that these documents privilege an economic rationality over other forms of human reason in standards that deal with our relationship with the rest of the world.[4] There are so many ways in which humans relate with the natural world. For instance, bauxite-rich Niyamgiri Mountain in the state of Chhattisgarh in India is venerated as a sacred site by the Kondh tribe that lives in that region. Similarly, the Standing Rock Sioux tribe treasure the land for spiritual and historical reasons on which the Dakota access pipeline is being built. In a slightly different vein, in 1872 the US Congress established the Yellowstone National Park and initiated a worldwide movement to convert open public lands to national parks and reserves so that people could enjoy the natural beauty of these lands. But such noneconomic ways of relating with the nonhuman world are relegated beyond the margins in the standards documents. Here the human-nature relationship is primarily represented as one in which nature either gets positioned as a resource that we ought to be sustainably exploiting for meeting our needs or as a recipient of harm that we cause to it on account of activities that are primarily economic in nature. For example, GSE high school earth science standard SES6 wants students to "Ask questions to investigate and communicate how humans depend on Earth's land and water resources" (refer to Appendix A: Referenced Science standard statements), and GPS sixth-grade standard S6E5 expects students to be able to describe "methods for conserving natural resources such as water, soil, and air" (refer to Appendix A: Referenced Science standard statements). In addition, in NGSS, though not in the Georgia documents, we find a positive valuation to adopting an economic criterion and by logical implication a marginalization of non-economic considerations in making decisions on the subject of natural resource management. For instance, consider the high school standard HS-ESS3-2: "Evaluate competing design solutions for developing, managing, and utilizing energy and mineral resources based on cost-benefit ratios" (refer to Appendix A: Referenced Science standard statements). Similarly, in the clarification statement accompanying standard HS-ESS3-3 that focuses on management of natural resources, only economic and technical factors, such as costs of resource extraction and waste management, per-capita consumption, and development of new technologies, are mentioned as affecting the management of natural resources (refer to Appendix A: Referenced Science standard statements).

The marginalization of the social and the collective dimension of human action and privileging of technocentric and economic values in science curricula shows that there is nothing objective about the representation of the world encoded in them. Contrary to NGSS's claim that science knowledge "indicates what can happen in natural systems not what should happen" (National Research Council, 2013; p. 6), we find that when it comes to human relations with the world, the science standards do encourage students to act in certain ways that can be very consequential for how they understand and engage with the rest of the planet. Because assumptions and implications remain tacit and unacknowledged, they can be quite persuasive in prodding unsuspecting students to adopt certain worldviews. In fact, they are seen as an important way to "manufacture consent" among the audience (Wodak, 2007). In the following section, we will see that the tacit assumptions and values in the science standards are closely tied with certain discourses that animate the standards documents.

Perspectives on the World

As we mentioned earlier, discourses as distinct forms of language in use are "ways of representing aspects of the world – the processes, relations and structures of the material world, the 'mental world' of thoughts, feelings, beliefs and so forth, and the social world" (Fairclough, 2003; p. 124). Of course, in addition to representation, discourses also have the key performative function of producing the world for us by establishing a regime of truth that makes certain statements about the world appear natural and universal, as in this is the way things *really* are, have always been, and will be. As we read the standards documents, it became apparent that these texts too are embedded in certain discourses that represent as well as produce a certain kind of world for students. In our textual analysis, we tried to identify the parts of the world that are presented or occluded in these documents and the points of view from which they are represented. In keeping with the interpretive framework of critical discourse analysis, our analysis was iterative and constant comparative as we continuously shifted back and forth and progressed from the ever-changing interpretive codes and themes. We found four major themes in these texts that pertained to (a) the ontological separation of the human and "natural" world, (b) ecosystem ecology-based conceptual framework, (c) the focus of the scientific gaze upon the world, and (d) technocentric-economic rationality in human-nature relationships. Further analysis coalesced these emergent themes into two interrelated discourses, a scientific discourse and an environmental discourse, that appeared to have the most influence in shaping these standards documents insofar as representations of the world and our relationships with it is concerned. A comingling of these mutually compatible discourses in science curricula contributes to the "greening" of school science (Veel, 2005) and fosters the construction of science students as apprentice environmentally responsible citizens. An elucidation of these discourses follows:

Scientific Discourse
Written by scientists and science educators, NGSS, GSE, and GPS present a Cartesian-Newtonian mechanistic perspective on the world that tells teachers and students that the world is out "there" as a physical reality that can be objectively observed, measured, modeled, and understood using scientific ideas and tools. The guiding document *A Framework for K-12 Science Education*, on which GSE and NGSS are based, recognizes that "science is fundamentally a social enterprise, and scientific knowledge advances through collaboration and in the context of a social system with well-developed norms" (National Research Council, 2012; p. 43). However, this acknowledgment gets lost in translation of guiding principles to actual science standards. Thus, we find that these standards expect students, as individuals, to carry out investigations, use computational thinking, develop and use scientific models to construct explanations (for science), and design solutions (for engineering). At the same time there is no recognition that alternative perspectives, methods, and interpretations of the world exist or are even possible.

Further, the scientific discourses in these standards documents divide the world into two distinct though interacting halves: a "natural world" comprising all the flora and fauna save humans along with the biophysical environment and a "social world" consisting of an undifferentiated and undivided collection of "humans" that depend on the "natural" world for their survival and well-being. This shows up in several content standards that expect students to understand how "humans" impact the natural world. For instance, in NGSS high school content standard HS-ESS3-4, we find that students are expected to "Evaluate or refine a technological solution that reduces impacts of human activities on natural systems" (refer to Appendix A: Referenced Science standard statements). Here it is evident that humans are seen as distinct and external to natural systems. Similarly, GSE high school content standard SEV4 expects students to "Obtain, evaluate, and communicate information to analyze human impact on natural resources" (refer to Appendix A: Referenced Science standard statements). The standards also show humans impacted by or dependent upon the natural world, such as in NGSS HS-ESS3-1 that focuses on constructing an explanation based on evidence that "the availability of natural resources, occurrence of natural hazards, and changes in climate have influenced human activity" (refer to Appendix A: Referenced Science standard statements).

Thus, exploration of the relationships between humans and the rest of the world is important in these standards documents. But in standards where these relationships are not the foci, learning science often gets reduced to learning about a hard-to-find pristine natural world that is unsullied by human presence or influence. For example, understanding relationships between organisms in ecosystems, movement of matter and energy in ecosystems, and processes that shape the Earth are important curricular emphases areas in the middle grades. On these topics the documents assume that we are talking about a planet where human presence can be safely ignored. For instance, in middle grades NGSS MS-ESS3-1 standard students are expected to learn "how the uneven distributions of Earth's mineral, energy, and groundwater resources are the result of past and current geoscience processes" (refer to Appendix A: Referenced Science standard statements). As is obvious, this standard represents a world that can be well understood without bringing humans and their activity into the picture. However, it is another matter altogether that ecologists now content that such a representation of the world does not correspond well with the world that most of us inhabit (Ellis & Ramankutty,

2008). Another example, from Georgia, would be the GSE standard S7 L4 in which students should "ask questions to gather and synthesize information from multiple sources to differentiate between Earth's major terrestrial biomes (i.e., tropical rain forest, savanna, temperate forest, desert, grassland, taiga, and tundra) and aquatic ecosystems (i.e., freshwater, estuaries, and marine)" (refer to Appendix A: Referenced Science standard statements). These biomes, as is also apparent from the examples of biomes in the GSE standard S7 L4, are traditionally defined only on the basis of the dominant plant form and climate and thus carry a tacit assumption of human absence.[5] Further, this way of parsing the planet into different regions of the world defined by climate and vegetation completely excludes the human-dominated anthropogenic biomes or anthromes, such as croplands and urban areas, that in fact cover more than 75% of Earth's ice-free land and account for 90% of terrestrial net primary production (Ellis & Ramankutty, 2008).

Finally, in keeping with the Cartesian-Newtonian orientation of the scientific discourse, the standards documents represent the world as a hierarchically organized and well-ordered system of interacting biotic and abiotic components that exchange matter and energy while performing different underlying processes that yield observable characteristics and phenomena. For example, GPS high school environmental science standard SEV2 wants students to be able to "demonstrate an understanding that the Earth is one interconnected system" (refer to Appendix A: Referenced Science standard statements). Such a representation of the world obviously carries a strong imprint of ecosystem ecology and Earth systems science (especially in the case of NGSS)—the two disciplinary areas of scientific inquiry based on a systems' view of the planet.

Thus, we find that in all the three standards documents, ecosystems are presented as the appropriate units for investigating and understanding the fundamental ecological processes and phenomena in the natural world. For instance, GSE fourth-grade standard S4 L1 expects students to be able to "Obtain, evaluate, and communicate information about the roles of organisms and the flow of energy within an ecosystem" (refer to Appendix A: Referenced Science standard statements). Similarly, NGSS middle-grade standard MS-LS2-1 expects students to "Analyze and interpret data to provide evidence for the effects of resource availability on organisms and populations of organisms in an ecosystem" (refer to Appendix A: Referenced Science standard statements). An ecosystem-based understanding of the natural world is especially noticeable in standards that relate to matter cycles and energy flows. Following the cybernetic systems perspective of ecosystem ecology, curriculum writers expect students to understand the relationships between organisms primarily in terms of cycling of matter and energy flows that occur between them. These exchanges are shown as occurring primarily within a bounded closed ecosystem and the exchange of matter and energy across the boundary with the world is ignored. Thus, we find standards like the GSE high school standard SEV1: "Obtain, evaluate, and communicate information to investigate the flow of energy and cycling of matter within an ecosystem" MS-LS2-1. It must be acknowledged here that perhaps because of the influence of Earth systems science, NGSS also has content standards that should lead students to understand energy flows and cycling of matter between larger Earth systems, such as hydrosphere, atmosphere, and biosphere. However, both in the case of ecosystems and of Earth systems, it is assumed that energy flows and matter cycles are not impacted by human activity. GSE also has some content standards that have been grouped under the topic category of "Earth Systems," but each of these content standards reflects traditional geology and Earth science topics and does not present Earth systems as a unit of analysis for understanding planetary phenomena and processes. GPS do not have any corresponding Earth science content standards.

We understand the pedagogic impulse to present a simplified representation of the world to students at the K-12 level. However, we worry if this also means that we are preparing students to accept an inaccurate and simplistic understanding of our world. This is not just because thermodynamically speaking ecosystems are perforce open systems that exchange matter and energy with the surrounding environment (Fath, 2014). But, more importantly, in today's world hardly any ecosystem has been left untouched by human impact (Chapin, Chapin, Matson, & Vitousek, 2011). As a result, human mediated outward flows of energy and matter from ecosystems in the form of "ecosystem services" and inward flows of energy and material pollution have rendered traditional notions of pristine and natural ecosystems moot (Gallagher & Carpenter, 1997). In fact, ecologists have begun to successfully argue that a "humans-free" ecosystem paradigm is not of much use in today's world because "leaving humans out of the ecological equation leads to inadequate explanations of ecosystem processes on an increasingly human-dominated Earth" (Marzluff et al., 2008; p. 148). Finally, despite the increasing recognition of the importance of organism and information flows in understanding ecosystems (Berkowitz, Nilon, & Hollweg, 2003; Cadenasso, Pickett, Weathers, & Jones, 2003; Ellis, 2015), the science standards continue to present ecosystem processes only in terms of matter and energy fluxes leading to another potential lacuna in students' understanding of our world.

In sum, it is evident that though ecosystem ecology has lost much of its shine and prestige as the leading paradigm among ecologists, it still remains the normative perspective in school science for understanding interactions among organisms and their environment. We need to seriously consider the continued relevance of the scientific discourse based on a mechanistic Cartesian-Newtonian perspective for science education that aims to prepare future citizens for life in the age of Anthropocene. As we show ahead, the scientific discourse embedded in these science standards meshes well and supports the other discourse—the environmental discourse—that we saw coursing through these documents. We see this close synergy between scientific and environmental discourses as not a matter of chance but rather as a reflection of the overall zeitgeist of our times that favors economically driven quantification and systematization of all aspects of our existence.

Environmental Discourse
A lay environmentalism has long been a staple of mainstream societal discourses in the United States (Kraft, 2015). As is currently the case with climate change, environmental issues are often at the center of vigorous debates and disputes both in public media and policy platforms. These debates are a reflection of a wide variety of perspectives and ways of talking about environmental issues that course through these dialogues. These perspectives and ways of talking about environmental issues have coalesced into a few key environmental discourses that have come to define the public and private dialogues, politics, and policymaking on issues related to our environment (Dryzek, 2013).

Societal issues often find their way into the school curriculum in the United States because of the longstanding tradition of seeing education as contributing to or responsible for solving social problems (Depaepe & Smeyers, 2008; Fendler, 2008). This happened with environmental issues as well. Environmental education is now seen as an important element of education, and there is an unmistakable "greening" of the school science in the United States (Veel, 2005). Thus, it was no surprise when our discourse analysis showed a clear evidence of an environmental discourse in the standards documents. Environmental discourses have been studied and classified in different ways in the research literature, see for instance Dryzek (2013). We found that the environmental discourse in school science curricula could not be neatly and satisfactorily classified into any of the pre-existing discursive categories as it is a hybrid discourse that borrows critical elements from two different existing discourses. We elucidate below how these two environmental discourses shaped the national and Georgia science standards.

Ecological Modernization
As mentioned earlier, there are two ways in which content standards present the natural world in school science—first, as a detached object of scientific inquiry through the scientific discourse as analyzed above and second, as the other relatum in nature-human relationships. It is in the later context that we find a heavy imprint of environmental discourses. One of these discourses is the discourse of ecological modernization (Dryzek, 2013). As a legitimizing discourse for capitalism in the age of environmental anxieties, it promises a "green" capitalism that offers both continued economic development and environmental sustainability. During critical discourse analysis this discourse revealed itself in the following ways. First, ecological modernization adopts a systems approach to conceptualize nature as a repository for meeting human needs. In the previous section we saw the clear preference for a systems approach in content standards to understanding planetary biophysical processes and phenomena. What was also unmistakable was the strong tendency in the standards documents to partition the natural world in terms of different resources. For example, NGSS informs science teachers and students that "Humans depend on Earth's land, ocean, atmosphere, and biosphere for many different resources" (MS-ESS3-1) and "resource availability has guided the development of human society" (HS-ESS3-1) (refer to Appendix A: Referenced Science standard statements). Thus, NGSS expects students to learn to "evaluate competing design solutions for developing, managing, and utilizing energy and mineral resources based on cost-benefit ratios" (HS-ESS3-2) and "create a computational simulation to illustrate the relationships among management of natural resources, the sustainability of human populations, and biodiversity" (HS-ESS3-3) (refer to Appendix A: Referenced Science standard statements). When students are not being asked to see the natural world as a repository of resources to exploit and manage, the natural world gets positioned as the

unfortunate recipient of damaging practices that we engage in to unsustainably exploit the natural world for our material purposes. Keeping this exploitative representation in mind, students are expected in the GSE standard (SEV4) to "Obtain, evaluate, and communicate information to analyze human impact on natural resources" and "Obtain, evaluate, and communicate information about the effects of human population growth on global ecosystems" (SEV5) (refer to Appendix A: Referenced Science standard statements).

When the natural world is presented as embedded in the human economy, this curricular choice pushes the diverse ways in which humans relate to and understand nature into the background and thus beyond the pale of official legitimation. For instance, we can ask why aesthetic or cultural dimensions of our relationship with the nonhuman world don't find a place in the science curricula while the economic dimension gets to hegemonize the curricular space. Dryzek (2013) notes that "ecological modernization pushes limits to growth into the background" (p. 145). Likewise, we did not find any acknowledgment in any standards document that continued economic growth based on exploitation of natural resources may not be ecologically possible and could pose a serious threat to our sustainable existence on this planet. Such an omission is remarkable in the face of the somber realization in the scientific community that there are indeed serious limits to continued economic growth especially now that we have already crossed three of the nine planetary boundaries within which we need to remain if we wish to avoid major calamitous environmental changes on a global scale (Rockstrom et al., 2009).[6]

Second, we found NGSS to also advocate a discourse of eco-managerialism (Luke, 1999) that tells students that natural resources can be and should be managed with the help of scientific knowledge and technological tools for meeting human needs because, after all, "the sustainability of human societies ... requires responsible management of natural resources" (HS-ESS3-3) (refer to Appendix A: Referenced Science standard statements). Thus, NGSS envisages students as evaluating "competing design solutions for developing, managing, and utilizing energy and mineral resources based on cost-benefit ratios" (HS-ESS3-2) and "creat[ing] a computational simulation to illustrate the relationships among management of natural resources, the sustainability of human populations, and biodiversity" (HS-ESS3-3) (refer to Appendix A: Referenced Science standard statements). Further, students are oriented to think that if environmental problems should arise while managing the natural world, technological solutions are always at hand to solve them. This optimism is guided by foundational beliefs that "scientists and engineers can make major contributions by developing technologies that produce less pollution and waste and that preclude ecosystem degradation" (HS-ESS3-4), and "though the magnitudes of human impacts are greater than they have ever been, so too are human abilities to model, predict, and manage current and future impacts" (ESS3.D) (refer to Appendix A: Referenced Science standard statements).

This extremely strong, positive assessment of the abilities of scientists and technologists to solve our environmental challenges even leads NGSS to suggest that "large-scale geoengineering design solutions (such as altering global temperatures by making large changes to the atmosphere or ocean)" could be appropriate ways to reduce "impacts of human activities on natural systems" (HS-ESS3-4) (refer to Appendix A: Referenced Science standard statements). This is despite the fact that many scientists think that geoengineering solutions are ethically problematic and rather than solving, in fact, may precipitate new ecological catastrophes (Gardiner, 2016; Shepherd, Iglesias-Rodriguez, & Yool, 2007). We are also struck by the complete marginalization of the sociopolitical dimensions of the environmental problems and their technological solutions in the NGSS. For instance, there is no recognition that local communities often possess precious ecological and sociocultural knowledge that is needed for a deeper and more complete understanding of environmental issues, and there is little hope for sustainable and just solutions of such problems without a democratic and equitable participation of all legitimate stakeholders (Sowman & Wynberg, 2014). Interestingly, Georgia science standards downplay eco-managerialism in their approach to nature-human relationships and focus instead on helping students understand how our use of natural resources impacts the natural world. We do not well understand why this is so, particularly given some of the history of eco-managerialism in the timber industry and other industries in Georgia, though it is at least partially related to the composition of the standards writing team. It certainly looks like an important and interesting question to explore in future research.

Third, in these documents we find that the technocratic, corporatist ways of managing the natural world are guided by an overall economic rationality in understanding the nature of environmental issues and weighing different possible solutions. Of course, the natural world is immediately brought within the purview of an economic logic as soon as the different components of the natural world are decontextualized from the local socioecological context and individuated as a resource or a sink. But NGSS then goes a step further by making economic criteria of cost-benefit ratios and efficiency the leading basis on which natural resource management decisions should be taken. Thus, we find content standards in which students are expected to "Evaluate competing design solutions for developing, managing, and utilizing energy and mineral resources based on cost-benefit ratios" (HS-ESS3-2), and consider "costs of resource extraction and waste management, per-capita consumption, and the development of new technologies" (Clarification statement accompanying HS-ESS3-3) as factors that affect the management of natural resources (refer to Appendix A: Referenced Science standard statements). Again, we don't see such an explicit legitimization of economic rationality in content standards of the state of Georgia.

In curricular matters, the null curricula of what is not taught is generally just as important as the intended curricula in laying out the learning possibilities for students (Flinders, Noddings, & Thornton, 1986). Thus, it becomes important to consider what alternate possibilities for conceptualizing relationships with the natural world are being lost when students are only exposed to a technocratic-economic logic. For instance, we worry that a precious opportunity has been squandered but not creating a curricular space that allows students to bring in a host of critical noneconomic factors, such as sociocultural traditions, socioeconomic justice, democratic participation, and ethics and aesthetics, in understanding and tackling environmental issues.

We note in passing that such curricular models that highlight sociocultural, socioeconomic, and sociolinguistic issues have been developed and implemented at different times in various school districts. One well-known, if sometimes controversial example, is the Portland Baseline Essay Project, developed and implemented in the Portland School District in Oregon. The goal of the baseline essay project was to integrate information about the history, culture, and contributions of local geocultural groups into the school curriculum across the content areas, including art, language arts, mathematics, science, social science, and music (Johnson & Williams, 2010). Importantly, local voices and perspectives, including those of indigenous and immigrant populations were reflected and highlighted in the curriculum. While the sometimes overly ambitious attempts to integrate marginalized topics, such as Afrocentric science, into the curriculum were critiqued (Travis, 1993), the baseline essay project served to infuse numerous aspects of what had been part of the null curriculum into the intended curriculum. As of 2017, these materials remain part of the Portland School District curriculum in some content areas, but with Oregon's adoption of the NGSS in 2014, they are no longer part of the district science curriculum.

Green Governmentality
The discourse of *ecological modernization* is pegged at a global scale as it aims to naturalize green capitalism as the only way to conceptualize our relationship with the natural world. However, it is not the only discourse that does this work. As Bäckstrand and Lövbrand (2006) assert, "Alongside the market-oriented approach to environmental problem-solving proposed by ecological modernization, a discourse of green governmentality predominates in industrialized societies" (p. 54). The discourse of green governmentality concurs with ecological modernization in its technocratic, managerial approach to nature and environmental threats. But its central focus is on the governance of individual and social life in matters related to their and society's relationship with the natural world. It is closely allied to or rather indistinguishable with neoliberalism in its conceptualization of individuals as autonomous, free, and responsible citizens who through their own volition choose to be prudent and responsible for their own destinies and choices. But while neoliberalism is broad and seeks to cover all aspects of our lives, green governmentality focuses on governing our conduct on issues related to the natural world. This governance is done through diverse "technologies of responsibilization" that individualize environmental responsibility and lead individuals to see themselves as primarily responsible through their individual acts for both trashing and rescuing the planet (Soneryd & Uggla, 2015).

This discourse reveals itself in the standards documents in the following ways. First, if it is not technological solutions, then local and individual environmental action is presented as the only other available option for preserving the environment and natural resources. This is truer for NGSS and GSE than GPS. For instance, in NGSS when students are expected to "evaluate competing design solutions for developing, managing, and utilizing energy and mineral resources based on cost-benefit ratios" (HS-ESS3-2), it is clarified that here the "emphasis is on the conservation, recycling, and reuse of resources" (refer to Appendix A: Referenced Science standard statements). Similarly, GSE high school standard (SEV5) envisage students as designing and defending "a sustainability plan to reduce your individual contribution to environmental impacts, taking into account how market forces and societal demands (including political, legal, social, and economic) influence personal choices" (refer to Appendix A: Referenced Science standard statements). Second, while backgrounding society's economic, political, and social systems as potential factors impacting nature-human relationships, students' attention is drawn towards population growth and individual consumption as the main reasons for the damage to the natural world. The NGSS middle-grade standard MS-ESS3-4 would be a good case in point: "Construct an argument supported by evidence for how increases in human population and per-capita consumption of natural resources impact Earth's systems" (refer to Appendix A: Referenced Science standard statements). Human population and per-capita consumption are indeed two very important direct drivers of anthropogenic impact on the natural world (Rosa, York, & Dietz, 2004). These two factors are included in the IPAT accounting equation that is widely used in understanding the anthropogenic ecological change. IPAT equation is written:

$$I_{mpacts} = P_{opulation} * A_{ffluence} * T_{echnology}$$

It is interesting that the standards documents completely occlude the third factor of "technology" in representing the anthropogenic causes of environmental issues. Here it is important to add that the third factor, "while labeled 'technology,' is really all other things, such as culture, institutional practices, and political processes" (Rosa, Rudel, York, Jorgenson, & Dietz, 2015). The backgrounding of the "technology" factor, thus, naturally works to strengthen the green governmentality discourse in these documents. The undemocratic governance structures and "green" capitalism are absolved of all responsibility in such a representation, while students are led to think that since it is they as individuals who are responsible for causing environmental damage, their environmental actions and choices (along with technological solutions) are needed to undo the harm. Interestingly, GPS is much weaker than the other two standards documents in its adherence to the discourse of green governmentality. While GPS implicates individuals as impacting the environment through their choices, it holds larger entities (businesses, governments, etc.) as equally responsible for impacting the environment by asking students to "Describe how energy and other resource utilization impact the environment and recognize that individuals as well as larger entities (businesses, governments, etc.) have impact on energy efficiency" (SEV4). Further, it counts population growth as an important factor but also expects students to understand "how political, legal, social, and economic decisions may affect global and local ecosystems" (SEV5). Considering that GPS are older standards and both NGSS and GSE are guided by the newer "A Framework for K-12 Science Education" (National Research Council, 2012), these differences can perhaps be seen as a plausible indication of how far the curricular discourse has moved towards neoliberal logic in the recent times.

Summing Up

Intended curricula can be seen as the crystallization of the core knowledge that the existing generation decides to pass on to the coming generation for the survival and continued prosperity of the society. As we saw in this chapter, the science content standards present one of the many possible representations of our world and the relationships we have with it. In these standards we see our world represented as a biophysical system that can be "terraformed" and sustainably managed by science and technology to support "green" capitalist societies on this planet. Further, these standards position us as environmentally responsible citizens who by our actions can doom or save the planet. Such a representation marginalizes the sociopolitical dimensions of the environmental problems and their technological solutions. This representation of our world has not only become dated from a scientific standpoint, it can also be seen as contributing to social injustice by delegitimizing local communities and many other groups with little power (such as the future inhabitants of the planet) as legitimate stakeholders in the decisions on what core knowledge of the world should be passed on to the future citizens. The science content standards do this by backgrounding their voices, knowledge, and experiences and misrepresenting their interests, representations, and values about how we should relate to the world around us.

In these standards we see a world that can be organized and run like a mechanized system to serve the material interests of those who have the means and positionality to access the goods and services it offers. Thus, after analyzing these standards we are left to question whether the national- or state-level science standards that will guide science teachers for at least the next decade present a picture of our world that benefits all of its existing and future inhabitants equally well. On the surface it appears that the writing of content standards has become a collaborative, dialogic, and democratic enterprise in the United States. The question then is how do we end up collaboratively writing science standards that are neither socially just nor ecologically wise. Attempts such as the Portland Baseline Essay Project, described earlier, remain outliers that seem even less likely to occur now, in the era of science standards adapted or adopted from the NGSS. Let us mull over this question for a while. We will return to it time and again in this book before venturing our best explanation in the last chapter.

NOTES

1. Because the standards documents are composed of short independent statements, we did not analyze *bridging assumptions* that give coherence to a text by logically linking different parts of a text.
2. *Georgia Performance Standards* had in turn replaced the earlier existing *Quality Core Curriculum*.
3. Standards writing process is a collaborative process involving many writers who may not always be on the same page regarding their understanding of science and curricular priorities. This can sometimes lead to inconsistencies within the standards. We found such a contradiction in a middle school NGSS standard MS-ESS3-1 which assumes that the humans have little, if any, role to play in uneven distribution of Earth's mineral, energy, and groundwater resources. This is clearly a problematic assumption as it distances school science from the current scientific view on the topic. However, the clarifying statement accompanying this standard goes against this assumption and states that in this standard the "Emphasis is on how these resources are limited and typically non-renewable, and how their distributions are significantly changing as a result of removal by humans." It is difficult to see how the same group of writers could have written both the standard and its accompanying clarification statement.
4. Here we understand economic rationality as an instrumental decision-making process that aims at finding the most efficient ways to reach ends that yield maximum returns with the least opportunity costs as measured in monetary terms.
5. For instance, a popular textbook on ecology, *Fundamentals of Ecology*, by Odum and Barrett (2005), defines biome as a "large regional or subcontinental system characterized by a particular major vegetation type (such as a temperate deciduous forest); biomes are distinguished by the predominant plants associated with a particular climate (especially temperature and precipitation)" (p. 513).

6. According to Rockstrom et al. (2009), the three planetary boundaries that we have already crossed are climate change, rate of biodiversity loss, and changes to the global nitrogen cycle.

REFERENCES

Anderson, M. (2017). *For Earth Day, here's how Americans view environmental issues*. Pew Research Center. Retrieved from http://www.pewresearch.org/fact-tank/2017/04/20/for-earth-day-heres-how-americans-view-environmental-issues/

Bäckstrand, K., & Lövbrand, E. (2006). Planting trees to mitigate climate change: Contested discourses of ecological modernization, green governmentality and civic environmentalism. *Global Environmental Politics, 6*(1), 50–75.

Berkowitz, A. R., Nilon, C. H., & Hollweg, K. S. (2003). *Understanding urban ecosystems: A new frontier for science and education*. New York, NY: Springer.

Berliner, D., & Biddle, B. (1995). *The manufactured crisis: Myths, fraud and the attack on America's public schools*. New York, NY: Perseus.

Cadenasso, M. L., Pickett, S. T., Weathers, K. C., & Jones, C. G. (2003). A framework for a theory of ecological boundaries. *AIBS Bulletin, 53*(8), 750–758.

Chapin, F. S., Chapin, M. C., Matson, P. A., & Vitousek, P. (2011). *Principles of terrestrial ecosystem ecology*. New York, NY: Springer.

Commission on Mathematics and Science Education. (2009). *Opportunity equation: Transforming mathematics and science education for citizenship and the global economy*. New York, NY: Carnegie Corporation of New York.

Committee on Development of an Addendum to the National Science Education Standards on Scientific Inquiry. (2000). *Inquiry and the national science education standards: A guide for teaching and learning*. Washington, DC: National Academy Press.

Depaepe, M., & Smeyers, P. (2008). Educationalization as an ongoing modernization process. *Educational Theory, 58*(4), 379–389.

Dryzek, J. S. (2013). *The politics of the Earth: Environmental discourses*. Oxford, UK: Oxford University Press.

Ellis, E. C. (2015). Ecology in an anthropogenic biosphere. *Ecological Monographs, 85*(3), 287–331. https://doi.org/10.1890/14-2274.1

Ellis, E. C., & Ramankutty, N. (2008). Putting people in the map: Anthropogenic biomes of the world. *Frontiers in Ecology and the Environment, 6*(8), 439–447. https://doi.org/10.1890/070062

Eriksen, T. H. (2014). *Globalization: The key concepts*. London, UK: A&C Black.

Fairclough, N. (1995). *Critical discourse analysis: Papers in the critical study of language*. Harlow, UK: Longman.

Fairclough, N. (2003). *Analysing discourse: Text analysis for social research*. London, UK: Routledge.

Fairclough, N. (2004). Critical discourse analysis in researching language in the new capitalism: Overdetermination, transdisciplinarity and textual analysis. In L. Young & C. Harrison (Eds.), *Systemic functional linguistics and critical discourse analysis* (pp. 103–122). London, UK: Continuum.
Fairclough, N., & Wodak, R. (2004). Critical discourse analysis. In T. A. van Dijk (Ed.), *Discourse as social interaction* (pp. 258–284). Thousands Oak, CA: Sage.
Fath, B. D. (2014). Ecosystem ecology. In S. E. Jorgensen & B. D. Fath (Eds.), *Encyclopedia of ecology* (pp. 1125–1131). Oxford, UK: Elsevier Science.
Fendler, L. (2008). New and improved educationalising: Faster, more powerful and longer lasting. *Ethics and Education, 3*(1), 15–26.
Flinders, D. J., Noddings, N., & Thornton, S. J. (1986). The null curriculum: Its theoretical basis and practical implications. *Curriculum Inquiry, 16*(1), 33–42.
Gallagher, R., & Carpenter, B. (1997). Human-dominated ecosystems. *Science, 277*(5325), 485–485.
Gardiner, S. M. (2016). Geoengineering: Ethical questions for deliberate climate manipulators. In S. M. Gardiner & A. Thompson (Eds.), *The Oxford handbook of environmental ethics*. Oxford, UK: Oxford University Press.
Georgia Department of Education. (n.d.). *Georgia performance standards for science*. Atlanta, GA.
Georgia Department of Education. (n.d.). *Georgia Standards of Excellence for Science*. Atlanta, GA.
Georgia Department of Education. (n.d.). *Georgia performance standards*. Retrieved from https://www.georgiastandards.org/Standards/Pages/BrowseStandards/BrowseGPS.aspx
Georgia Department of Education: Science. (n.d.). Retrieved from http://www.gadoe.org/Curriculum-Instruction-and-Assessment/Curriculum-and-Instruction/Pages/Science.aspx.
Georgia Science Teachers Association. (n.d.). *Science standards for Georgia's next generation*. Retrieved from http://www.georgiascienceteacher.org/Next-Gen-Updates
Goertz, M. E. (2009). Standards-based reform: Lessons from the past, directions for the future. In K. K. Wong & R. Rothman (Eds.), *Clio at the table: Using history to inform and improve education policy* (pp. 201–219). New York, NY: Peter Lang.
Goertz, M. E. (2010). National standards: Lessons from the past, directions for the future. In B. J. Reys & R. E. Reys (Eds.), *Mathematics curriculum: Issues, trends, and future directions: 2010 yearbook* (pp. 51–63). Reston, VA: National Council of Teachers of Mathematics.
How to read the next generation science standards. (2013). Retrieved from https://www.nextgenscience.org/sites/default/files/How%20to%20Read%20NGSS%20-%20Final%2008.19.13.pdf

Hufnagel, E., Kelly, G. J., & Henderson, J. A. (2017). How the environment is positioned in the Next Generation Science Standards: A critical discourse analysis. *Environmental Education Research*, 1–23. https://doi.org/10.1080/13504622.2017.1334876

Johnson, E., & Williams, F. (2010). Desegregation and multiculturalism in the Portland public schools. *Oregon Historical Quarterly, 111*(1), 6–37.

Kraft, M. (2015). *Environmental policy and politics*. New York, NY: Taylor & Francis.

Luke, T. (1999). Eco-Managerialism. Environmental Studies as a Power/Knowledge Formation. In F. Fischer & M. A. Hajer (Eds.), *Living with nature. Environmental politics as cultural discourse* (pp. 103–120). Oxford, UK: Oxford University Press.

Macrine, S. L., McLaren, P., & Hill, D. (2010). *Revolutionizing pedagogy: Education for social justice within and beyond global neo-liberalism*. New York, NY: Palgrave Macmillan.

Martin, J. R., & Rose, D. (2008). *Genre relations: Mapping culture*. London, UK: Equinox Pub.

Marzluff, J., Shulenberger, E., Endlicher, W., Alberti, M., Bradley, G., Ryan, C., ... Simon, U. (2008). *Urban ecology: An international perspective on the interaction between humans and nature*. New York, NY: Springer US.

Mazid, B. M. (2014). *CDA and PDA made simple: Language, ideology and power in politics and media*. Newcastle, UK: Cambridge Scholars Publisher.

National Research Council. (2001). *Investigating the influence of standards: A framework for research in mathematics, science, and technology education*. Washington, DC: The National Academies Press.

National Research Council. (2012). *A framework for K-12 science education: Practices, crosscutting concepts, and core ideas*. Washington, DC: National Academies Press.

National Research Council. (2013). *Next generation science standards: For states by states*. Retrieved from https://www.nextgenscience.org/

Nelson, F. (Ed.). (2012). *Community rights, conservation and contested land: The politics of natural resource governance in Africa*. New York, NY: Routledge.

Odum, E. P., & Barrett, G. W. (2005). *Fundamentals of ecology*. Belmont, CA: Thomson Brooks/Cole.

Pinar, W. F. (2012). *What is curriculum theory?* London, UK: Routledge.

Pruitt, S. L. (2014). The next generation science standards: The features and challenges. *Journal of Science Teacher Education, 25*(2), 145–156.

Ravitch, D. (2010). *The death and life of the great American school system: How testing and choice are undermining education*. New York, NY: Basic Books.

Rockström, J., Steffen, W., Noone, K., Persson, Å., Chapin, F. S., III, Lambin, E., ... Schellnhuber, H. J. (2009). Planetary boundaries: Exploring the safe operating space for humanity. *Ecology and Society, 14*(2).

Rosa, E. A., Rudel, T. K., York, R., Jorgenson, A. K., & Dietz, T. (2015). The human (anthropogenic) driving forces of global climate change. In R. E. Dunlap & R. J. Brulle (Eds.), *Climate change and society* (pp. 32–60). New York, NY: Oxford University Press.

Rosa, E. A., York, R., & Dietz, T. (2004). Tracking the anthropogenic drivers of ecological impacts. *Ambio: A Journal of the Human Environment, 33*(8), 509–512.

Sharma, A. (2016). STEM-ification of education: The zombie reform strikes again. *Journal for Activist Science and Technology Education, 7*(1), 42–51.

Shepard, L. A. (2015). If we know so much from research on learning, why are educational reforms not successful? In M. J. Feuer, A. I. Berman, & R. C. Atkinson (Eds.), *Past as prologue* (p. 41). Washington, DC: National Academy of Education.

Shepherd, J., Iglesias-Rodriguez, D., & Yool, A. (2007). Geo-engineering might cause, not cure, problems. *Nature, 449*(7164), 781–781.

Soneryd, L., & Uggla, Y. (2015). Green governmentality and responsibilization: New forms of governance and responses to 'consumer responsibility'. *Environmental Politics, 24*(6), 913–931. https://doi.org/10.1080/09644016.2015.1055885

Sowman, M., & Wynberg, R. (2014). *Governance for justice and environmental sustainability: Lessons across natural resource sectors in Sub-Saharan Africa*. New York, NY: Taylor & Francis.

The need for standards. (n.d.). Retrieved from https://www.nextgenscience.org/need-standards

The next generation science standards: Executive summary. (2013). Retrieved from https://www.nextgenscience.org/sites/default/files/Final%20Release%20NGSS%20Front%20Matter%20-%206.17.13%20Update_0.pdf

Tran, D., Reys, B. J., Teuscher, D., Dingman, S., & Kasmer, L. (2016). Analysis of curriculum standards: An important research area. *Journal for Research in Mathematics Education, 47*(2), 118–133.

Travis, J. (1993). Schools stumble on an Afrocentric science essay. *Science, 262*(5136), 1121–1123.

Veel, R. (2005). The greening of school science. In A. P. L. J. R. Martin, J. R. Martin, & R. Veel (Eds.), *Reading science: Critical and functional perspectives on discourses of science* (pp. 115–151). New York, NY: Taylor & Francis.

Wixson, K. K., Dutro, E., & Athan, R. G. (2003). Chapter 3: The challenge of developing content standards. *Review of Research in Education, 27*(1), 69–107.

Wodak, R. (2007). Pragmatics and critical discourse analysis: A cross-disciplinary inquiry. *Pragmatics & Cognition, 15*(1), 203–225.

Writing Team. (n.d.). Retrieved from https://nextgenscience.org/writing-team

CHAPTER 4

The Intended Curriculum: Nature as Represented in a Science Textbook

A lay environmentalism has become part of mainstream American culture and values (Kempton, Boster, & Hartley, 1995; Sellers, 2012). However, available evidence suggests that most Americans are poorly equipped with the knowledge necessary for informed environmental action (Lorenzoni & Pidgeon, 2006; Sterman & Sweeney, 2007). Furthermore, research also indicates that as currently practiced, academic instruction is not adequately preparing K-12 and college students to perceive natural and social systems as fundamentally coupled or to understand the nature of the imbalance in current human-natural systems (Assaraf & Damri, 2009; Covitt, Tan, Tsurusaki, & Anderson, 2009). Our failure to equip students with such an understanding of the world is likely to play an important role in how they perceive environmental issues and their importance later in life.

Of course, students' attitudes about and understanding of environmental issues have diverse provenances. As participants in multiple local, global, and "glocal" cultural/social contexts, young people actively and passively imbibe knowledge, ideas, attitudes, folklore, and perspectives about the environment and their relationship with it from a vast array of sources (Eagles & Demare, 1999; Weaver, 2002). However, school

---

This chapter is a reprint of Sharma, A., & Buxton, C. A. (2015). Human–Nature Relationships in School Science: A Critical Discourse Analysis of a Middle-Grade Science Textbook. *Science Education, 99*(2), 260–281.

© The Author(s) 2018
A. Sharma, C. Buxton, *The Natural World and Science Education in the United States*, https://doi.org/10.1007/978-3-319-76186-2_4

science, as a carrier of official, authoritative knowledge, constitutes one of the critical formative influences (Tikka, Kuitunen, & Tynys, 2000). This authoritative knowledge is coded in science textbooks sanctioned for use by states and local school districts. Thus, it can be said that science textbooks present the official interpretation of the relationship between humans and the world they live in and draw sustenance from.

In this study, we explore how the language of a science textbook works to represent environmental problems and solutions in distinct ways that may have serious implications for students' ecological literacy. Specifically, and in recognition to the importance of microlevel analysis of texts to reveal global discourses, we investigate a seventh-grade textbook, *Georgia: Holt Science and Technology: Life Science* (Allen, Berg, Christopher, Duschek, & Taylor, 2008) to answer the following research questions:

1. How are humans textually represented as interacting with natural systems in this science textbook?
2. What is the nature of environmental problems and solutions as textually represented in the textbook?

In these research questions, the distinction between and characterization of some systems as "natural" and some as "social" has been done for purely heuristic reasons, as we found that this distinction facilitated clearer analysis. We acknowledge that because of the pervasive and deep influence of human activity in all biomes of this planet, natural and human systems have become closely coupled and are best understood as intrinsically socioecological (Ellis, Klein Goldewijk, Siebert, Lightman, & Ramankutty, 2010; Liu et al., 2007).

Our research questions and the decision to investigate one science textbook in depth are based on a few key considerations. First, studies have long shown that student learning is strongly influenced by textbooks (Ball & Feiman-Nemser, 1988; Kelly, 2007; Tyson, 1997; Weiss, Pasley, Smith, Banilower, & Heck, 2003). As Kelly (2007) noted, "the significance of the textbook in classroom learning extends beyond its direct influence on a student's comprehension of the subject matter, as it typically serves as a guideline for instructional choices and for the sequence of learning events" (p. 459). Second, school science presents only a selective representation of science as understood and practiced by scientists, and this selected representation is greatly shaped by extant social and political contexts (DeBoer, 1991; Rudolph, 2002, 2003). Thus, researchers need to investigate the preferred "official" representation of science in school so as to better understand how school science is shaped by broader sociopolitical forces. Science textbooks stand out as the most promising place to begin this work. In fact, we believe that a careful analysis of science textbooks can serve as a politically engaged critique of the official discourse of school science that can reveal the much-needed generative possibilities for a more democratic and justice-oriented school science. While there has been a clear trend among education researchers on understanding discourse as reflected in speech, such as interactional classroom talk (Kelly, 2007; Rogers, Malancharuvil-Berkes, Mosley, Hui, & Joseph, 2005), a similar emphasis on examining written texts, such as science textbooks, as conduits of authoritative discourses has been lacking in research.

Finally, ecological issues, such as global climate change, are primarily societal issues that directly pertain to unsustainable usage of environmental products and services in global and local frameworks of ecological governance that are largely iniquitous and undemocratic (Sharma, 2012). However, barring some research on K-12 as well as college students' understandings of matter cycles, water cycles, coupling of human and natural systems, and individual decision-making about environmental issues (Assaraf & Damri, 2009; Menzel & Bogeholz, 2009; Mohan, Chen, & Anderson, 2009), we found that there is scant research on the relationships among contexts, practices, and representations of natural and social systems and individuals in science curriculum material and classroom instruction. The research reported here seeks to partially fill this gap and to provide direction for further studies of these relationships.

Here it is important to acknowledge that even though analysis of science texts continues to be a peripheral issue in science education research, there are a few researchers who have systematically analyzed texts used in science classrooms from critical discourse and systemic functional linguistics (SFL) perspectives, highlighting the nature of official school science discourse (Carlone & Webb, 2006; Fang, 2005, 2006; Hanrahan, 2006; Schleppegrell, 2004). This important body of research has tended to focus on examining the school science discourse as a semiotic system with distinct syntax and semantics to understand the linguistic challenges science texts pose for students in terms of reading and comprehension. For instance, Fang (2005, 2006) has analyzed key linguistic features of science texts, such as nominalization of grammar and widespread use of subordinate clauses, propositions, conjunctions, and pronouns, to illuminate the language demands of science reading in schools. We share these concerns and have in fact written about them (Sharma & Anderson, 2009). The key issues in this article, however, are more specifically focused on the nature of knowledge about the natural world that a particular science textbook seeks to convey and not on how difficult it is for students to comprehend that knowledge.

The focus of this article speaks to the ability of texts to serve certain social and discursive functions through ontological representation and production of our material and social worlds. Thus, we have chosen to investigate our research questions through a theoretical lens that borrows heavily from Fairclough's (2004) dialectical-relational approach to critical discourse analysis (CDA). Fairclough's approach, in turn, selectively appropriates analytic tools from SFL for textual analysis. In the next section, we pull together the theoretical ideas and concepts from CDA and SFL to build the theoretical framework for our study.

THEORETICAL FRAMEWORK

This study follows the discursive turn in social science research by accepting the idea that discourse is central to understanding social life in advanced capitalist societies (Erickson, 2004; Fairclough, 2003). As a result, we have adopted a critical-discourse-analysis-based theoretical framework because it can account for the function of texts in social practices, such as teaching and learning of science. A core premise of this framework is that all social practices, including science learning, involve meaning making by participants through language and other forms, such as visual images and body language. That is, "every practice has a semiotic element" (Fairclough, 2004, p. 122). In Fairclough's textually oriented approach to critical discourse analysis (TODA), the focus is on those semiotic elements of social practice that are marked by distinctly identifiable patterns of language in use, in speech as well as writing (Fairclough & Wodak, 2004). Discourse in TODA is taken to mean a distinct form of language in use as an element of social practice.

Our article focuses on the representational aspect of the official school science discourse. Therefore, we have analyzed our chosen science textbook at the microlevel (semiotic text analysis) and mesolevel (content text analysis). Fairclough's approach is appropriate for this purpose as it yokes social theory with textual analysis. That is, if following Fairclough and Wodak (2004) we assume that discourses constitute social reality and shape our representations of ourselves and of the world, it becomes imperative for us to look closely at language use in these representations to understand this process. For this kind of close-grained textual analysis, TODA is able to borrow analytical tools from SFL as they share a common assumption that language is primarily functional in nature.

As opposed to a structural linguistic focus on how texts are composed, such as in Saussure's approach to linguistics, SFL is geared towards understanding what language does. Thus, the key assumption of SFL is that "language has evolved to satisfy human needs; and the way it is organized is functional with respect to these needs – it is not arbitrary" (Halliday, 1994, p. xiii). Thus, SFL presents and analyzes "language as shaped (even in its grammar) by the social functions it has come to serve" (Fairclough, 2004, p. 126). According to Halliday (1994) language serves three broad "metafunctions." First, it helps us understand the world (both material and social) by enabling textual representation of our experience of the world (ideational metafunction). Second, language allows us to interact with other participants (interpersonal metafunction). Third, by organizing and structuring linguistic information for cohesive and coherent text production, language helps us create communicative and meaningful messages (textual metafunction).

According to SFL, these metafunctions are accomplished by the grammar of the language as the grammar is primarily functional in nature, "in the sense that everything in it can be explained, ultimately, by reference to how language is used" (Halliday, 1994, p. xiii). Grammar serves the ideational metafunction of making human experience of the world meaningful. It also helps us enact our interpersonal relationships by "sharing experiences with others, giving orders, making offers and so on" (Halliday, 2004, p. 9). For our purposes, the important thing to note is that a specific language-in-use discourse in our framework has embedded within it a distinct stance on the world. In this way, our article presents a middle-grade science textbook's stance on the world.

In the past decade or so, educational researchers have turned to CDA to explore important issues at the intersections of language and society in school settings (Rogers et al., 2005). Though limited and still developing, CDA- and SFL-based research on school science has also grown in the past decade (Carlone & Webb, 2006; Fang, 2005, 2006; Hanrahan, 2006; Schleppegrell, 2004). Our study aims to contribute to this small but growing pool of research in science education.

METHODS

The ability of dominant, global discourses to seep through local boundaries and influence local representations, practices, and institutions has been explored by many scholars, such as Bourdieu and Wacquant (2001) and Rose (1999). Our study shows how such seepage happens in one specific albeit important instance of the representation of the natural world and its relationship with the social world in a science textbook. In this regard, our study follows a well-established trail in CDA where in-depth, single-case-study investigations have shown how discourses do their work in representing and producing the world we live in (e.g., refer to de los Heros, 2009; Dennis, 2011; Merkl-Davies & Koller, 2012). Thus, it is our hope that once readers are able to see how the theorized "universal" particularizes itself in one context through naturalist generalization (Stake, 1995) they will be able to transfer, adapt, and apply explanations emerging from our study to their own and other contexts.

The analytic framework of CDA rests on an analysis of "dialectical relationships between semiosis (including language) and other elements of social practices" (Fairclough, 2004, p. 123). As indicated earlier, because the semiotic aspect of school science textbooks has been shown to have a powerful impact on the teaching and learning of school science qua social practices, it was appropriate that we centered our study on the textual analysis of a science textbook. We chose the *Georgia: Holt Science and Technology: Life Science* (Allen et al., 2008) seventh-grade science textbook for our analysis. This book was selected because it is in seventh grade that students in Georgia are first introduced to fundamental ecology concepts in their science instruction, such as ecosystems, biodiversity, and matter and energy cycles, which are essential for understanding the relationship between social and natural worlds.

In the state of Georgia, the textbooks recommended for use in public schools are selected on the basis of recommendations of the State Textbook Advisory Committee. This committee organizes evaluations of all textbooks submitted by publishers of record. It consists of professional educators and members of the community who are appointed from each congressional district and the state at large. The committee serves in an advisory capacity, and final adoption decisions are made by the state board of education. (http://www.doe.k12.ga.us/Curriculum-Instruction-and-Assessment/Curriculum-and-Instruction/Pages/

Learning-Resources.aspx). The textbook we selected for analysis is on the Georgia state recommended list of K-12 science learning resources (http://www.doe.k12.ga.us/Curriculum-Instruction-and-Assessment/Curriculum-and-Instruction/Pages/Learning-Resources.aspx) and is currently being used as the official textbook in the school district where one of us is currently doing ethnographic research on classroom discourse in a middle-grade science classroom. It is also in use in many other school districts in the state of Georgia. The text has an ecology unit comprising four chapters—Interactions of Living Things (chapter 18, pp. 478–505), Cycles in Nature (chapter 19, pp. 506–523), Earth's Ecosystems (chapter 20, pp. 524–551), and Environment Problems and Solutions (chapter 21, pp. 552–575). For a middle-grade student, these chapters constitute her first systematic introduction to an ecological perspective on the living world. This type of purposeful sampling of school science text allowed us to illustrate a typical case of how science textbooks represent the world for the students.

As is true for all modern science textbooks, the text we analyzed offers a wide variety of format and content. Each of the analyzed chapters begins with a list of relevant Georgia Performance Standards, a prereading activity, and a start-up activity. The main textual body of each chapter consists of sections and subtopics written as short paragraphs. Alongside the main body of the chapter, there are text boxes on the margins that give definitions of terms, suggestions for mathematical practice, Internet activities, quick lab activities, and connections to language arts. At the beginning of each section, there is a box that indicates what students will learn, key section vocabulary, and a reading strategy to practice. Diagrams or figures with captions also support the main textual body of each chapter. The chapters end with additional supplements, such as a "Science Skills Activity," "Skills Practice Lab," "Chapter Review," and "CRCT (Criterion-Referenced Competency Test) Preparation."[1]

Thus, this textbook is multimodal in the sense of having different linguistic, visual, and spatial modes and designs. In this study, we focused only on the main textual body comprising the relevant chapter sections. Of course, an analysis of other components of a textbook can be insightful, as has been shown by the analysis of visual representations by Lee (2010) and Potter and Rosser (1992). However, we have found in an ongoing classroom-based research by one of us (Sharma, 2013) in local, middle-grade science classrooms that science teachers give much more attention to the main body text than they do to other components of the chapter. Thus, a detailed and systematic analysis of the linguistic elements of the main body text can be assumed to reveal much about the representations of the natural and social world presented to teachers and students by science textbooks. At a pragmatic level, it is also important to note that SFL offers a fine-grained analysis of text done at the clausal level. Taken together, the main body of the four chapters we analyzed consists of 1191 clauses spread over 185 paragraphs. This amounted to analyzing a pool of data big enough to lead us to believe that we had reached the data saturation limit.

We analyzed the textual data of each chapter at two levels: content (mesolevel) and semiotic (microlevel). First, taking a paragraph as a unit of communication and analysis, we did a mesolevel coarse-grained, emergent qualitative document analysis (Altheide, Coyle, DeVriese, & Schneider, 2010). After several rounds of close reading of the chapters, we identified the theme of each paragraph (Martin, 1992). Then through an iterative process of constant comparison and coding of paragraph themes, we identified a few broad overarching themes characterizing the representation of relationship between the natural and social world and the role of human beings as individuals in that relationship (Emerson, Fretz, & Shaw, 1995; Fairclough & Wodak, 2004).

Next, to understand the representational resources of the text, that is, "the productive and innovative potential of language as a meaning-making system" (The New London Group, 1996, p. 79), we did a SFL-based, fine-grained microlevel analysis of all the clauses—independent as well as dependent—in the main textual body. In SFL, the clause is chosen as the unit of analysis because it is the smallest grammatical unit that can convey a meaningful message in a text. Because we were interested in understanding how school science texts represent the world for the students, this round of analysis primarily focused on uncovering the ideational metafunction of the selected text through transitivity analysis, specifying how "phenomena of the real world are represented as linguistic structures" (Halliday, 1994, p. 102). The other two metafunctions, textual and interpersonal, of the selected text were not analyzed as they were not relevant to our research questions, which focused only on the nature of representation of the natural world in the textbook.

An analysis of transitivity was helpful in understanding how individuals have been positioned in representation of the relationship between the social and natural world. As The New London Group (1996) points out, "transitivity indicates how much agency and effect one designs into a sentence" (p. 79). They further argue that "Since we humans connect agency and effect with responsibility and blame in many domains (discourses), these are not just matters of grammar. They are ways of designing language to engage in actions like blaming, avoiding blame, or backgrounding certain things against others" (p. 80). These issues of agency and effect became prominent in our analysis.

According to Halliday (1994),

> the basic semantic framework for the representation of processes is very simple. A process potentially consists of three components: (i) the process itself (realized by verbal groups); (ii) participants in the process (realized by nominal groups); (iii) circumstances associated with the process (realized by prepositional phrases or adverbials that indicate time, place or manner). These provide the frame of reference for interpreting our experience of what goes on. (p. 101)

Analyzing each clause, we identified each of these three process components. Our criteria for identification and categorization of processes, participants, and circumstances are summarized below.

Processes

Material Processes. Processes that reflected some kind of "doing." These clauses usually describe concrete, tangible actions that can be done by any kind of entity—both living (plants, animals, human beings, etc.) and nonliving (water, air, etc.).

Mental Processes. Processes that indicated cognition (thinking, knowing, understanding, etc.), affection (liking, fearing, etc.), and perception (seeing, hearing, etc.).

Relational Processes. Processes that reflect distinct ways of being. For instance, following Halliday (1994), "x is a" was identified as an intensive relational process, whereas "x has a" was called a possessive relational process. We also identified circumstantial relational processes of the type "x is at a."

Existential Processes. Processes that "represent experience by positing that 'there was/is something'" (Eggins, 2004, p. 238).

Participants in Processes

For each clause, we first identified the different participants in the process. Then, we checked if the active participants (i.e., those in agentive roles), such as actors in material clauses, were explicitly mentioned or not in the clause. If they were, we made a further determination if the reference was:

- generic (Ge) in nature, such as in people, and humans;
- to a group (Gr), as in farmers, ranchers, scientists, companies, etc.;
- to a specific individual or entity (S), such as author, the Environmental Protection Agency and the United States Fish and Wildlife Service.

Second, if the active participant was not explicitly mentioned in the clause, then we decided if this exclusion was a case of:

Exclusion backgrounding (EB): Where the identity of the active participant can be inferred from the immediate textual context. For example, "Farmers often use chemicals to control fungi. Growing other plants among the bananas, or increasing biodiversity, can also prevent the spread of fungi" (p. 564). In the underlined clause, we find that even though the actors have not been included in the clause, we can infer from the preceding sentence that the clause is referring to farmers as actors.

Exclusion suppression (ES): Where the identity of the active participant is excluded and suppressed so that it cannot be inferred even from the preceding and succeeding text. For instance, "CFCs (chlorofluorocarbons) were used in aerosols, refrigerators, and plastics" (p. 555).

Circumstances Associated with Processes

Here we noted the circumstances, if mentioned, in which the process described in the clause took place. Science as a body of knowledge about the world has a tendency of abstracting phenomena and processes from a specific time and place and coding them in a temporally and spatially independent manner (Halliday & Martin, 1993). This is of obvious importance to students' understanding of science as a body of knowledge that can be leveraged to understand the world around them. Therefore, we categorized the circumstance of the process in each clause as time definite or indefinite and as space definite or indefinite.

The next section presents the results of our analysis. However, before we proceed it is important to acknowledge the ineluctable interpretive nature of our analysis. While doing our analysis, time and again we found ourselves in agreement with Halliday's (1994) assertion that "an exegetical work of this kind, whether ideological, literary, educational, or anything else, is a work of interpretation" (p. xvi). Our analysis is also explanatory in nature. This is because the functional orientation of an SFL-based analysis is able to not only reveal what a text means but also and equally important it explains how a text means (Eggins, 2004). In the case of our study, therefore, such an analysis helped us understand how the text in question positions the relationship between the natural and social worlds and the role of individuals in this relationship.

Results

Guided by the research questions, our analysis led us to results that showed us how text can function as a form of technology to construct representations that serve particular sociopolitical purposes. We came to better appreciate the ways grammar can be used to elide or muddy the role of humans in creating and sustaining environmental problems. Our analysis also afforded us a deeper understanding of how school science texts can function to represent a sanitized version of environmental threats to students and lead them towards certain preferred ways to respond to such threats. Combining emerging results from mesolevel qualitative documents analysis with microlevel transitivity analysis of the text through an iterative process of constant comparison and multistage coding led to three overarching themes that we present in this section: obfuscating human agency in human-nature interactions, externalization of environmental threat, and the individualization of environmental action.

Obfuscating Human Agency in Human-Nature Interactions

As mentioned earlier, the ecology unit in the textbook we analyzed comprises four chapters. The first three chapters are aimed at presenting the key ideas and topics of ecology. These chapters offer an introductory portrait of Earth as a natural system. The fourth chapter is focused on presenting the natural-social relationship through a survey of important environmental problems and solutions. Main topics covered in these chapters are given in Table 4.1.

Table 4.1 Chapter outlines

Chapter 18: Interactions of Living Things	Chapter 19: Cycles in Nature	Chapter 20: The Earth's Ecosystems	Chapter 21: Environmental Problems and Solutions
Section 1: Everything is connected. Biotic and abiotic parts of an environment Levels of organization in the environment Populations Communities Ecosystems The biosphere Section 2: Living things need energy. Producers, consumers and decomposers Food chains and food webs Energy pyramids Balance in ecosystems Section 3: Types of interactions. Interactions with the environment Interactions between organisms Competition Predators and prey Symbiosis Coevolution	Section 1: The cycles of matter. The water cycle The carbon cycle The nitrogen cycle Many (other) cycles Section 2: Ecological succession. Regrowth of a forest Primary succession Secondary succession Mature communities and biodiversity	Section 1: Land biomes. The Earth's land biomes Forests Grasslands Deserts Tundra Section 2: Marine ecosystems. Life in the ocean Temperature as a factor in marine ecosystems Different zones according to depth and sunlight in marine ecosystems Ecosystems near shoreline Section 3: Freshwater ecosystems Stream and river ecosystems Pond and lake ecosystems Wetland ecosystems How a lake or a wetland can become a forest?	Section 1: Environmental problems. Pollution Resource depletion Exotic species Human population growth Habitat destruction Effects on humans Section 2: Environmental solutions Conservation Reduce Reuse Recycle Maintaining biodiversity Environmental strategies

Studying the "Other" World We were interested in finding out how the textbook represents the relationship between social and natural systems. So, we began by identifying clauses that involved the social world or human beings in any way in all clause-type categories (see Table 4.2). In the first three chapters that focus on presenting an ecological account of the natural world, a vast majority of the clauses make no reference to

THE INTENDED CURRICULUM: NATURE AS REPRESENTED IN A SCIENCE... 99

human beings or the social world, the only exception being the mental process clauses that largely refer to humans thinking or feeling about some issues related to the natural world. In these chapters, most of the material clauses that showed human engagement with the natural world represented humans as scientists studying the natural systems. For example, consider the following statement on p. 483:

Ecologists	study	the biosphere	to learn	how	organisms	interact	with the abiotic environment
Actor	Process: material	Goal	Circumstance		Actor	Process: material	Goal

⬅━━━━━━━━━━━━━━━━━━━━━➡ ⬅━━━━━━━━━━━━━━━➡
            Clause 1   Clause 2

Thus, in the chapters that present an ecological perspective on the natural world, the textbook textually creates a representation of a "pristine" natural world in which human presence is marginal and mostly limited to scientific investigations of natural systems. For instance, the third chapter, The Earth's Ecosystems, divides the Earth into one marine biome and various land biomes, such as forests, grasslands, and deserts. Although ecologists (see, for instance, Alessa & Chapin, 2008; Ellis & Ramankutty, 2008) now consider anthropogenic biomes[2] as the most important biomes on the planet, these biomes are not even mentioned in

Table 4.2 Clauses involving humans/social world

Chapter	Material process clause	Relational process clause	Existential process clause	Mental process clause	Total
18—Interactions of Living Things	17	0	0	19	36 (of 353)
19—Cycles in Nature	11	0	0	4	15 (of 125)
20—Earth's Ecosystems	25	4	0	16	45 (of 358)
21—Environmental Problems and Solutions	179	90	8	22	299 (of 355)
Total	232	94	8	61	395 (of 1191)

the textbook. Furthermore, in none of the biomes mentioned in the third chapter do we find human beings or human societies mentioned as members of those biomes or any of their ecosystems. Other animals and plants are, of course, frequently mentioned.

As a result, even in the case of ecosystems that are heavily impacted by human actions, the textbook scrubs off all human references and represents phenomena occurring in them purely as natural phenomena. For instance, consider the following paragraph on lakes and ponds in the chapter "The Earth's Ecosystems":

> Did you know that a lake or pond can disappear? How can this happen? Water entering a standing body of water usually carries nutrients and sediment. These materials settle to the bottom of the pond of lake. Dead leaves from overhanging trees and decaying plant and animal life also settle to the bottom. Then, bacteria decompose this material. This process uses oxygen in the water. The loss of oxygen affects the kinds of animals that can survive in the pond or lake. For example, many fishes would not be able to survive with less oxygen in the water. (p. 543)

This text describes the degradation or death of lakes through the process of eutrophication. Though eutrophication can happen both from human-caused and natural factors, in the current world, this process happens mostly because of human-induced factors (Ansari, Gill, Lanza, & Rast, 2010). By presenting eutrophication as a completely natural phenomenon, the text obfuscates our culpability in the ongoing rampant degradation of water bodies all over the world.

Erasing and Impersonalizing Human Culpability in Environmental Problems We wished to investigate how humans, individually and/or collectively, are positioned in the natural-social relationship constructed in the textbook. Therefore, we next looked at the clauses that referred to the natural-social interaction of a material nature. Here we were interested in interactions that were beyond investigations of natural systems by scientists. We first examined material process clauses in all four chapters as these clauses reflect some kind of "doing" and usually have an actor to initiate and sustain the process. The results of our analysis are summarized in Table 4.3.

As Table 4.3 indicates, the textbook authors show a remarkable, persistent tendency to exclude and suppress human agency in material clauses by not only failing to mention any actor but also by making it impossible to guess the identity of the actor from the context. To show how this is done, consider the following example on p. 488 from the textbook:

Once common throughout much of the United States	gray wolves	were almost wiped out		as the wilderness		was settled	
Circumstance	Process: material	Goal		Actor: ES (excluded and suppressed)	Goal	Process: material	Actor: ES (excluded and suppressed)

⬅————— Clause 1 —————➡   ⬅————— Clause 2 —————➡

This sentence comprises two material clauses that speak about a distinctly material process of the near extinction of gray wolves, but neither clause mentions any actor. The immediate context of this clause—the paragraph—does not give any clue either. So, the grammar of the clause works to divert students' attention away from the human agency involved in causing this situation and focus it on the process instead.

Another example from p. 558:

Tropical rain forests,	(are)	the most diverse habitats on Earth	are sometimes cleared	for farmlands, roads and lumber.		
Carrier	Process: Relational	Attribute	Process: Material	Goal		Actor: ES (excluded and suppressed)

⬅————— Clause 1 —————➡   ⬅————— Clause 2 —————➡

Here too we see the same exclusion and suppression of human agency as one never gets to know who clears these tropical rain forests. The paragraph in which this clause is imbedded is of no help either. From Table 4.3, one can see that human agency is excluded and suppressed in 44% of the clauses that deal with natural-social interaction.

In the remaining occurrences, actors are mentioned, but usually in the most generic terms, such as "people" and "humans." For instance, the topic of "Habitat Destruction" in the chapter "Environmental Problems and Solutions" starts on p. 558 as follows:

Table 4.3 Agency analysis in material process clauses

Chapter	Generic reference (# of clauses)	Group reference (# of clauses)	Specific reference (# of clauses)	Exclusion and suppression (# of clauses)	Total (# of clauses)
18—Interactions of Living Things	0	0	0	5	5
19—Cycles in Nature	0	2	0	2	4
20—Earth's Ecosystems	1	0	0	0	1
21—Environment Problems and Solutions	46	23	5	52	126
Total	47 (35%)	25 (19%)	5 (4%)	59 (44%)	133

People	need	Homes.	People	also need	food and building materials.
Carrier	Process: Relational	Attribute	Carrier	Process: Relational	Attribute

⟵——— Clause 1 ———⟶ ⟵——— Clause 2 ———⟶

But	when land is cleared	for construction, crops, mines, or lumber,		the topsoil	may erode
	Process: material	Goal		Actor: ES (excluded and suppressed)	Actor	Process: material

⟵——— Clause 3 ———⟶ ⟵——— Clause 4 ———⟶

Today	lumber companies	often plant	new trees	to replace the trees	that were cut down	
Circumstance	Actor: Gr (group)	Process: material	Goal	Circumstance	Process: material	Actor: ES (excluded and suppressed)

⟵——— Clause 1 ———⟶ ⟵——— Clause 2 ———⟶

In this entire section, the only human actors held responsible for habitat destruction are "people" or "humans." Lumber companies—a more specific group of actors—are mentioned once on p. 558, and in a positive light:

By pinning the agency for habitat destruction on "people," the textbook makes it appear as if ordinary laypersons are largely culpable for habitat destruction. This is a dated and incorrect view given the evidence that in the past few decades much of the habitat loss has occurred not because of the activities of "people," but through globalized exploitation of land and natural resources by corporations (Laurance, 2010; Rudel, Defries, Asner, & Laurance, 2009). Such a linguistic move on the part of the authors works to misrepresent the nature and causes of habitat destruction on our planet. In this and several other similar instances, we must conclude that by labeling actors generically as "people" or "humans" in ecological processes and phenomena involving interactions between natural and social systems, the textbook obfuscates the nature of human involvement with natural systems.

Furthermore, we found that the textbook presents a largely consumption-focused and individual-oriented perspective on environmental problems. Our thematic analysis showed that the net impact of such a skewed presentation is to cast environmental problems as matters of excessive resource consumption and ecologically harmful actions, such as improper waste disposal and spreading invasive species, by individuals (mostly as members of some, amorphous and anonymous group labeled as "people").

By elevating resource consumption by "people" as the sole cause of environmental problems, the textbook oversimplifies the complexities of natural-social interactions and elides more influential and larger sociocultural and politico-economic factors behind environmental stress and degradation (Hempel, 1996). For instance, the chapter on environmental problems and solutions does not even mention the much more influential globalized, industrialized system of production and distribution of ecosystem-based goods and services or what has been called the treadmill of production by Schnaiberg, Pellow, and Weinberg (2000) in which individual resource consumption takes place. The chapter "Environmental Problems and Solutions" does mention industrial production as a source of pollution once, but relegates it to the past by declaring: "today, machines don't produce as much pollution as they once did. But there are more sources of pollution today than there once were" (p. 554). The remainder of the chapter elaborates on these other putatively current sources of environmental problems, with actions taken by "people" completely eclipsing industrial production and distribution as causal factors of environmental

People	need	(many chemicals)	(people)	and use	many chemicals.
Carrier: Generic (Ge)	Process: Relational	Attribute	Actor: Generic; exclusion backgrounded (Ge/EB)	Process: material	Goal

⟵——— Clause 1 ———⟶ ⟵——— Clause 2 ———⟶

Some chemicals	are used	(by people)	to treat diseases
Goal	Process: Material	Actor: Generic; exclusion backgrounded (Ge/EB)	Circumstance

⟵——— Clause 3 ———⟶

Other chemicals	are used	(by people)	in plastics and preserved foods.	Sometimes	the same chemicals	that help	people
Goal		Actor: Generic; exclusion backgrounded (Ge/EB)	Circumstance	Cirumstance	Actor: Group (Gr)	Process: material	Goal

⟵——— Clause 4 ———⟶ ⟵——— Clause 5 ———⟶

(the same chemical)	may harm	the environment.
Actor: Group; exclusion backgrounded (Gr/EB)	Process: material	Goal

⟵——— Clause 6 ———⟶

problems. For instance, while discussing chemicals as a source of pollution, the text on p. 555 mentions:

Contrary to what the textbook implies, production activities such as farming, industry, and mining, rather than the end-usage by people, are generally considered to be the major sources of chemicals-based pollution (McKinney, Schoch, & Yonavjak, 2012; Miller & Spoolman, 2007). As a result, we see such a representation of chemicals-based pollution as pulling a linguistic smokescreen over the more culpable actors of this kind of pollution. Our paragraph-based thematic analysis of the selected text allowed us to notice that all references to the private sector, such as companies engaged in resource extraction, were positive, whereas people figured both as harming and helping the environment (see Table 4.4). In light of studies, such as Freudenburg (2005) and Grant, Bergesen, and Jones (2002), that have shown how socially structured and iniquitous environmental and discursive privileges allow many companies to disproportionately pollute and damage natural systems, such a one-sided representation of environmental problems appears highly questionable.

Externalization of Environmental Threat

In consonance with the representation of natural systems and social systems as separate and independent from each other, the textbook presents animals and plants, but not humans, as the most likely victims of environmental problems. For instance, the chapter "Environmental Problems and Solutions" presents habitat destruction as an environmental problem and illustrates it with the following description of nonpoint-source water pollution:

> A second kind of water pollution is nonpoint-source pollution. This kind of pollution comes from many different sources. Nonpoint-source pollution often happens when chemicals on land are washed into rivers, lakes, and oceans. These chemicals can harm or kill many of the organisms that live in marine habitats. (p. 558)

This paragraph is typical of the entire section on habitat destruction in which there is no mention about the impact of this problem on humans. The chapter "Environmental Problems and Solutions" includes only two paragraphs in which effects of environmental problems on humans are discussed. In these instances, the worst that environmental problems seem

Table 4.4 Positive and negative references to major actors in environmental problems and solutions

Chapter	Good guys					Bad guys			
	Scientists	Communities	EPA[a]	People	Companies	People	Farmers	Companies
18—Interactions of Living Things	2	0	0	0	0	0	0	0
19—Cycles in Nature	1	0	0	0	0	0	0	0
20—Earth's Ecosystems	1	0	0	0	0	0	0	0
21—Environment Problems and Solutions	7	2	1	4	3	3	1	1
Total	10	2	1	4	3	3	1	1

[a]EPA—The United States Environmental Protection Agency

to be doing to humans is to make them sick. There is a brief mention of one other (and more severe) consequence—depletion of natural resources. But that is presented as a tentative future challenge that succeeding generations may have to face.

We find it surprising that the textbook takes no cognizance of the fact that we already live in a world where environmental problems pose grave threats to livelihoods and physical survival in many human communities around the world (Boano, Zetter, & Morris, 2007; Leighton, Shen, Warner, & Zissener, 2011). As a result of environmental stress and degradation, every year many people in poor, developing nations are compelled to leave their homes and join the fast growing hordes of environmental refugees all over the world (Myers, 2002; Westra, 2009). It is estimated that by 2020 there may be as many as 50 million environmental refugees on this planet (Zeitvogel, 2011). Few Americans face such threats, and the global system of resource usage and waste disposal insulates most of them from its ill effects. It is quite probable that most Americans do not feel concerned about environmental issues affecting marginal communities in distant places given that a lack of direct sensory information about environmental problems and poor access to mediated information on environmental issues are two major barriers to environmental concern among people (Takacs-Santa, 2007). Opinion polls among Americans also do not register environmental problems facing poor communities in developing nations as environmental concerns (see, for instance, Saad, 2011). Thus, it seems to us that by presenting a uniquely privileged (and first-world) perspective on environmental problems that hides the true human costs of the environmental problems, the textbook can be seen as abetting the status quo in terms of environmental perceptions among Americans. In fact, in an ongoing ethnographic study in a suburban school in the state of Georgia in which one of us is currently engaged, we have indeed found that most students do not see human beings as being harmed by environmental problems in any major way and view these problems largely as threats to animals and plants.

Finally, we also found it remarkable that Chapter 21, "Environmental Problems and Solutions," does not mention climate change as an environmental problem even though the entire scientific community (barring a few contrarians, of course) views it as a grave threat to the planet (G8+5, 2009). Without naming climate change or global warming, the text does mention the likely consequences of an increase in atmospheric carbon dioxide, but this is presented as a case of gaseous pollution. Furthermore, the causal link between the increase in atmospheric carbon dioxide and an increase in global temperatures is presented not as a conclusion on which there is near universal scientific consensus, but merely as an opinion of "many scientists." It is important to keep in mind that we are talking here about a 2008 publication, and thus it cannot be presumed that the fundamentals of climate science were not settled when this book was written or revised.

The Individualization of Environmental Action

As the primary solution to environmental problems, the chapter "Environmental Problems and Solutions" suggests conservation through reducing, reusing, and recycling of resources by individual members of society. It also enjoins students and citizens to help maintain biodiversity by petitioning the government to put threatened species on the federal government's endangered species list. Such actions are indeed worthwhile for addressing many of our environmental problems. However, we see a mismatching of ecological and sociological scales between the problems and suggested solutions. First, the scope of environmental action suggested in the chapter is essentially local, whereas all of the problems mentioned in this chapter, such as pollution, resource depletion, exotic species, overpopulation, and habitat destruction, exist on local as well as larger regional, national, and international scales—both socially and ecologically. Local action is necessary, but rarely sufficient in solving such problems (Ostrom, Burger, Field, Norgaard, & Policansky, 1999). The chapter does not give any hint that large-scale solutions, such as international treaties on marine fishing and restrictions on use of CFCs in manufacturing, may also be needed to effectively solve the environmental problems enumerated in the chapter. In this sense, the text fails to prepare students to understand either the complexities of the environmental problems or the inadequacies of the suggested solutions.

The kind of environmental actions recommended by the textbook, such as conservation of natural resources through reduction, reuse, and recycling, are all actions recommended for individuals. Because environmental problems are presented in the text as being largely caused by individual action, the suggested solutions appear commensurate with the problems. They are also in agreement with the mainstream environmentalist discourse that tends to present environmental action as personal virtue (Treanor, 2010). This discourse implicitly assumes that an individual is an autonomous agent in sole possession of her agency, and her carbon footprint reflects her freely chosen lifestyle. In light of research findings (see, for instance, Lorenzoni, Nicholson-Cole, & Whitmarsh, 2007) that show that there exist significant social barriers in modern societies that inhibit even knowledgeable and environmentally conscious people from acting in environmentally responsible ways, we find such assumptions highly questionable.

Furthermore, in agreement with most ecologists and environmental sociologists (e.g., see Dietz, Ostrom, & Stern 2003; Hempel, 1996), we believe that individual environmental action can only be effective when it is situated in the context of just and democratic governance of ecological resources. In the absence of fair and democratic management of ecological commons, individuals have more to lose and little to gain through reduce, reuse, and recycle types of virtuous environmental actions—a situation that invariably leads to unsustainable exploitation of common property resources resulting in what has been called the tragedy of commons (Hardin, 2009). Therefore, by presenting environmental action primarily as an individual virtue, the textbook misses a precious opportunity to educate students about the larger sociopolitical contexts that are needed for such actions to be effective in meeting current environmental challenges.

These results, interpretive in nature as they are, reflect our partial understanding of the text from a CDA- and SFL-oriented perspective. The full import of these results can only be appreciated with the context of usage in middle school science classrooms in mind and when connected with wider research literature and societal discourses. This is a task we seek to accomplish in the final section.

Discussion

Guided by a belief in the ability of texts to perform an important social function of ontologically representing and discursively producing our material and social worlds, we undertook the study reported in this article to better understand how one typical middle-grade science textbook may be shaping students' understanding of their relationship to social systems. We were keen to find out how individuals were textually represented in the relationships between natural and social systems. Finally, while exploring these questions we also wondered how these representations might be connected with dominant societal discourses.

We found that the grammar of the text has been so shaped as to either exclude and suppress human agency in natural-social relationships or to attribute that agency to some anonymous, amorphous, and nonindividuated group, labeled simply as "people" or "humans." Thus, the overall impression given to teachers and students is that while it is important for them to understand environmental phenomena, it is not particularly relevant to know the specifics of human involvement in creating, modifying, or sustaining those phenomena. Furthermore, by attributing the agency for creating or aggravating environmental problems to "people," the textbook makes it appear as if ordinary laypersons are largely culpable for these problems.

We also found the representation of environmental problems and solutions in the textbook to be simplistic, one-sided, and not very helpful in preparing students to deal with the environmental problems described in the book. We were particularly struck by the omission of climate change as an important environmental problem in its own right. Furthermore, the textbook attempts to sanitize environmental problems by omitting any mention of the devastating impact environmental degradation is having on marginalized communities in other nations. Ordinary people are presented as culprits for causing environmental problems through excessive consumption and ecologically damaging actions, such as transporting invasive species. Additionally, we found that the textbook offers local, individual-based solutions to global, societal ecological problems. When taken together, these features of the textbook serve to completely elide the role of industrialized manufacturing and distribution systems and agencies involved in it, while presenting a sociological and ecological mismatch in scale between problems and proposed solutions.

The school science discourse reflected in the text clearly borrows some important features from the broader discourse of science and technology. As a result, it presents a perspective on the world that is much different from that of everyday discourses. For instance, the suppression and impersonalization of human agency is a common feature of science and technology discourses (Bernard & Philip, 2000)—also a notable feature in the selected textbook.

We have been critical of the representation of nature in this textbook, yet such a representation is quite emblematic of how the broader discipline of ecology viewed nature until quite recently. As Bradshaw and Bekoff (2001) explain, it was only when global environmental problems escalated and ecology got drawn into the social arena that ecologists realized that the traditional separation of the natural and the social world was no longer helpful and the ecological embeddedness of the social world needed to be a common starting point for ecological research. Rather than labeling the textbook's representation of the natural world as untrue, it is better seen as dated and in urgent need of revision so that it reflects the current state of ecological understanding. Thus, one implication of our findings is a call for a closer integration of school science and social studies, especially on ecological topics. While the national social studies curricular framework has included the integration of science, technology, and society as one of ten core themes for nearly two decades (National Council for the Social Studies, 1994, 2010), the integration of science and social studies still remains quite limited in the latest conceptual framework for science education (National Research Council [NRC], 2011; see also Feinstein & Kirchgasler, 2015). If we wish for students in more developed nations to be better informed about the effects of their actions on marginalized communities around the world, we will need a broadening of the privileged, first-world perspective of textbooks to include the concerns and perspectives of less privileged communities.

Environmental change can be viewed from multiple perspectives, including Schnaiberg's (1980) "treadmill of production" model that implicates capitalism in degradation of the environment, Bunker's (1990) use of world systems theory to explain environmental issues, and the IPAT equation model, according to which, environmental impact is a function of population, affluence, and technology (I (environmental impact) = P (population) × A (affluence) × T (technology)) (Ehrlich & Ehrlich, 1992). Similarly, Hempel (1996) has identified multiple driving forces for environmental problems, such as core societal values, amplifiers (population growth and technology), consumptive behavior, and political economy. From among the multiple perspectives and driving forces these varied models provide, the textbook we analyzed appears to significantly refer to just one factor, affluence, and just one driving force—consumptive behavior.

We understand that all textbook authors must make decisions about simplifying and limiting the scope of content to ensure that material is comprehensible. One might argue that seventh-grade students cannot be expected to understand environmental phenomena from a multiplicity of perspectives, and hence a simplistic perspective of individual-based consumptive behavior is the most pedagogically advisable option. We do not agree with such a view, believing that by the seventh grade, most students are intellectually and emotionally mature enough to appreciate and benefit from complex perspectives of life around them (Kuhn, 2005).

Rather, we suspect that the authors' choices reveal the productive power of national policy discourses on science and education as well as other dominant societal discourses that shape influential texts. We have analyzed the middle-grade science standards of all 50 states of the United States (Sharma & Buxton, 2012) and found remarkable similarities between the analyzed textbook and state science standards in the way the relationship between the social and natural world is conceptualized and presented. Furthermore, it is our view that the latest conceptual framework for science education (NRC, 2011) does not do much to challenge the status quo in this regard. Textbook adoption is a political process with significant economic implications for the publishers (Finn & Ravitch, 2004). In such a situation, textbook authors and their publishers may have a strong incentive to publish books that are reflective of the dominant societal discourses and that fail to challenge the status quo. However, unless science education equips students with worldviews, knowledge, and practices that enable them to change the unsustainable state of social-nature relationships, we do not see how we can be hopeful about our common future.

Finally, we will claim that in addition to the dominant science and technology discourse, the discourse of neoliberalism also appears to be influencing authors and publishers of the current generation of school science textbooks. This discourse has become naturalized in most industrialized societies and is now the dominant way to conceptualize relationships between individuals, society, and nature (Himley, 2008). Neoliberalism endorses a robust individualism within societies, in the sense of seeing individual actions as reflective of people's choices rather than as outcomes of sociocultural contexts, a perspective that venerates market-based exchanges for circulation of goods and services (Harvey, 2005). Echoes of this discourse can be clearly seen in the way the textbook neatly separates the social from the natural world. This dis-embedded view of the social world prepares students as future citizens to accept the commodification of nature as unproblematic. It also naturalizes the metabolic rift, that is, the material distancing of human beings in modern societies from the natural conditions of their existence in school science discourse (Foster, 1999). We can see the influence of neoliberalism in the textbook's attribution of environmental problems to individual action and the promotion of environmental solutions as a personal virtue. In view of severe environmental challenges, such as climate change, that clearly require collective social action, we need current and future citizens who act as sharing, cooperative, and public good maximizing homo reciprocans (Fehr & Gintis, 2007), rather than as self-interested, rational, and utility maximizing homo economicus (Peters, 2001). By highlighting, through detailed discourse analysis, how one middle school science textbook represents the ecological relationships between natural and social systems, this study argues for reformed science textbooks that bring the preparation of such citizens closer to reality.

To better understand science, Halliday and Martin (1993) urged science teachers to "work towards a much clearer grasp of the function of language as technology in building up a scientific picture of the world" (p. 202). Our study supports this recommendation. It is clear to us that science teachers (and teacher educators) will be well served in their work by gaining a better understanding of how texts (whether textbooks, the news, or other communications about science for the general public) position the environment, climate, and a range of actors through particular linguistic practices with critical sociocultural implications. Furthermore, research seems to suggest that greater (though selective) usage of science trade books can help teachers offer better explanations to scientific phenomena and counter dominant narratives with alternative perspectives (Fang, Lamme, & Pringle, 2010; Smolkin, McTigue, Donovan, & Coleman, 2009). Thus, our study urges science teachers to look beyond prescribed textbooks for curricular support. If the teacher understands well how deeply the social world is embedded with the natural world and knows where to look, then it should be possible for her to find appropriate supplementary science texts and other curricular resources that counter the dominant mainstream discourses impacting the teaching of basic ecology in schools. It is likely that most science teacher preparation programs are not doing enough to prepare teachers for this task (Zak & Munson, 2008). Thus, a re-envisioning of courses for both preservice and in-service teachers of science may be required. In light of urgent threats from global climate change and other environmental crises that have resulted from the unsustainable and socioecologically unjust construction of our relationship with the natural world, it is our hope that studies such as our CDA of a commonly used seventh-grade life science textbook will contribute to making ecological issues and closer integration with social studies a key concern in science education research, professional learning, and curriculum development.

NOTES

1. A list of questions, most of them multiple-choice, to prepare students for the state end-of-year test.
2. Also called anthromes or human biomes.

REFERENCES

Alessa, L., & Chapin, F. S. (2008). Anthropogenic biomes: A key contribution to earth-system science. *Trends in Ecology & Evolution, 23*(10), 529–531. https://doi.org/10.1016/j.tree.2008.07.002

Allen, K., Berg, L., Christopher, B., Dushek, J., & Taylor, M. (2008). *Georgia: Holt science and technology: Life science*. Austin, TX: Holt, Rinehart and Winston.

Altheide, D., Coyle, M., DeVriese, K., & Schneider, C. (2010). Emergent qualitative document analysis. In S. N. Hesse-Biber & P. Leavy (Eds.), *Handbook of emergent methods* (pp. 127–151). New York, NY: Guilford Publications.

Ansari, A. A., Gill, S. S., Lanza, G. R., & Rast, W. (2010). *Eutrophication: Causes, consequences and control*. New York, NY: Springer.

Assaraf, O. B.-Z., & Damri, S. (2009). University science graduates' environmental perceptions regarding industry. *Journal of Science Education and Technology, 18*(5), 367–381.

Ball, D. L., & Feiman-Nemser, S. (1988). Using textbooks and teachers' guides: A dilemma for beginning teachers and teacher educators. *Curriculum Inquiry, 18*(4), 401–423.

Bernard, J. M., & Philip, G. (2000). Technocratic discourse: A primer. *Journal of Technical Writing and Communication, 30*(3), 223–251.

Boano, C., Zetter, R., & Morris, T. (2007). *Environmentally displaced people: Understanding the linkages between environmental change, livelihoods and forced migration*. Oxford, UK: University of Oxford, Department of International Development.

Bourdieu, P., & Wacquant, L. (2001). New liberal speak: Notes on the new planetary vulgate. *Radical Philosophy*, (105). Retrieved from http://www.radicalphilosophy.com/default.asp?channel_id=2187&editorial_id=9956

Bradshaw, G. A., & Bekoff, M. (2001). Ecology and social responsibility: The re-embodiment of science. *Trends in Ecology and Evolution, 16*(8), 460–465. https://doi.org/10.1016/s0169-5347(01)02204-2

Bunker, S. G. (1990). *Underdeveloping the Amazon: Extraction, unequal exchange, and the failure of the modern state*. Chicago, IL: University of Chicago Press.

Carlone, H. B., & Webb, S. M. (2006). On (not) overcoming our history of hierarchy: Complexities of university/school collaboration. *Science Education, 90*(3), 544–568.

Covitt, B. A., Tan, E., Tsurusaki, B. K., & Anderson, C. W. (2009). *Students' use of scientific knowledge and practices when making decisions in citizens' roles.* Paper presented at the annual conference of the National Association for Research in Science Teaching from http://edr1.educ.msu.edu/EnvironmentalLit/publicsite/html/report_2009.html

de los Heros, S. (2009). Linguistic pluralism or prescriptivism? A CDA of language ideologies in "Talento," Peru's official textbook for the first-year of high school. *Linguistics and Education: An International Research Journal, 20*(2), 172–199.

DeBoer, G. E. (1991). *A history of ideas in science education: Implications for practice.* New York, NY: Teachers College Press.

Dennis, C. (2011). Measuring quality, framing what we know: A critical discourse analysis of the common inspection framework. *Literacy, 45*(3), 119–125.

Dietz, T., Ostrom, E., & Stern, P. C. (2003). The struggle to govern the commons. *Science, 302*(5652), 1907–1912. https://doi.org/10.1126/science.1091015

Eagles, P. F. J., & Demare, R. (1999). Factors influencing children's environmental attitudes. *The Journal of Environmental Education, 30*(4), 33–37. https://doi.org/10.1080/00958969909601882

Eggins, S. (2004). *An introduction to systemic functional linguistics.* New York, NY: Continuum.

Ehrlich, P. R., & Ehrlich, A. H. (1992). *Healing the planet.* Sydney, NSW: Surrey Beatty & Sons.

Ellis, E. C., Klein Goldewijk, K., Siebert, S., Lightman, D., & Ramankutty, N. (2010). Anthropogenic transformation of the biomes, 1700 to 2000. *Global Ecology and Biogeography, 19*(5), 589–606. https://doi.org/10.1111/j.1466-8238.2010.00540.x

Ellis, E. C., & Ramankutty, N. (2008). Putting people in the map: Anthropogenic biomes of the world. *Frontiers in Ecology and the Environment, 6*(8), 439–447. https://doi.org/10.1890/070062

Emerson, R. M., Fretz, R. I., & Shaw, L. L. (1995). *Writing ethnographic fieldnotes.* Chicago, IL: The University of Chicago Press.

Erickson, F. (2004). *Talk and social theory.* Malden, MA: Polity Press.

Fairclough, N. (2003). *Analysing discourse: Text analysis for social research.* London, UK: Routledge.

Fairclough, N. (2004). Critical discourse analysis as a method in social scientific research. In R. Wodak & M. Meyer (Eds.), *Methods of critical discourse analysis* (pp. 121–138). Thousand Oaks, CA: Sage.

Fairclough, N., & Wodak, R. (2004). Critical discourse analysis. In T. A. van Dijk (Ed.), *Discourse as social interaction* (pp. 258–284). Thousand Oaks, CA: Sage.

Fang, Z. (2005). Scientific literacy: A systemic functional linguistics perspective. *Science Education, 89*(2), 335–347.

Fang, Z. (2006). The language demands of science reading in middle school. *International Journal of Science Education, 28*(5), 491–520.
Fang, Z., Lamme, L. L., & Pringle, R. M. (2010). *Language and literacy in inquiry-based science classrooms, grades 3-8.* Thousand Oaks, CA: Sage.
Fehr, E., & Gintis, H. (2007). Human motivation and social cooperation: Experimental and analytical foundations. *Annual Review of Sociology, 33*(1), 43–64. https://doi.org/10.1146/annurev.soc.33.040406.131812
Feinstein, N., & Kirchgasler, K. (2015). Sustainability in science education? How the next generation science standards approach sustainability, and why it matters. *Science Education, 99*(1), 121–144.
Finn, C. E., & Ravitch, D. (2004). *The mad, mad world of textbook adoption.* Retrieved from http://www.edexcellencemedia.net/publications/2004/200409_madworldoftextbookadoption/Mad%20World_Test2.pdf
Foster, J. B. (1999). Marx's theory of metabolic rift: Classical foundations for environmental sociology. *American Journal of Sociology, 105*(2), 366.
Freudenburg, W. R. (2005). Privileged access, privileged accounts: Toward a socially structured theory of resources and discourses. *Social Forces, 84*(1), 89–114.
G8+5. (2009). *G8+5 Academies' joint statement: Climate change and the transformation of energy technologies for a low carbon future.* Retrieved from www.nationalacademies.org/includes/G8+5energy-climate09.pdf
Grant, D. S., Bergesen, A. J., & Jones, A. W. (2002). Organizational size and pollution: The case of the U.S. chemical industry. *American Sociological Review, 67*(3), 389–407.
Halliday, M. (1994). *An introduction to functional grammar.* London, UK: Hodder Arnold.
Halliday, M. (2004). *The language of science* (Vol. 5). New York, NY: Continuum.
Halliday, M., & Martin, J. (1993). *Writing science: Literacy and discursive power.* Pittsburgh, PA: University of Pittsburgh.
Hanrahan, M. U. (2006). Highlighting hybridity: A critical discourse analysis of teacher talk in science classrooms. *Science Education, 90*(1), 8–43.
Hardin, G. (2009). The tragedy of the commons. *Journal of Natural Resources Policy Research, 1*(3), 243–253.
Harvey, D. (2005). *A brief history of neoliberalism.* New York, NY: Oxford University Press.
Hempel, L. C. (1996). *Environmental governance: The global challenge.* Washington, DC: Island Press.
Himley, M. (2008). Geographies of environmental governance: The nexus of nature and neoliberalism. *Geography Compass, 2*(2), 433–451. https://doi.org/10.1111/j.1749-8198.2008.00094.x

Kelly, G. J. (2007). Discourse in science classrooms. In S. K. Abell & N. G. Lederman (Eds.), *Handbook of research on science education* (pp. 443–470). Mahwah, NJ: Lawrence Erlbaum.

Kempton, W., Boster, J. S., & Hartley, J. A. (1995). *Environmental values and American culture.* Cambridge, MA: MIT Press.

Kuhn, D. (2005). *Education for thinking.* Cambridge, MA: Harvard University Press.

Laurance, W. F. (2010). Habitat destruction: Death by a thousand cuts. In N. S. Sodhi & P. R. Ehrlich (Eds.), *Conservation biology for all* (pp. 73–86). New York, NY: Oxford University Press.

Lee, V. R. (2010). Adaptations and continuities in the use and design of visual representations in U.S. middle school science textbooks. *International Journal of Science Education, 32*(8), 1099–1126.

Leighton, M., Shen, X., Warner, K., & Zissener, M. (2011). *Policy and institutional mechanisms to address the needs of climate-related migrants.* Retrieved from http://i.unu.edu/media/unu.edu/publication/000/027/540/Research-Brief-3.pdf

Liu, J., Dietz, T., Carpenter, S., Alberti, M., Folke, C., Moran, E., ... Taylor, W. (2007). Complexity of coupled human and natural systems. *Science, 317*(5844), 1513–1516. https://doi.org/10.1126/science.1144004

Lorenzoni, I., Nicholson-Cole, S., & Whitmarsh, L. (2007). Barriers perceived to engaging with climate change among the UK public and their policy implications. *Global Environmental Change Part A: Human & Policy Dimensions, 17*(3/4), 445–459.

Lorenzoni, I., & Pidgeon, N. (2006). Public views on climate change: European and USA perspectives. *Climatic Change, 77*(1), 73–95.

Martin, J. R. (1992). *English text: System and structure.* Philadelphia, PA: John Benjamins Publishing Company.

McKinney, M. L., Schoch, R. M., & Yonavjak, L. (2012). *Environmental science: Systems and solutions.* Burlington, MA: Jones & Bartlett Learning, LLC.

Menzel, S., & Bogeholz, S. (2009). The loss of biodiversity as a challenge for sustainable development: How do pupils in Chile and Germany perceive resource dilemmas? *Research in Science Education, 39*(4), 429–447.

Merkl-Davies, D. M., & Koller, V. (2012). 'Metaphoring' people out of this world: A critical discourse analysis of a chairman's statement of a UK defence firm. *Accounting Forum, 36*(3), 178–193. https://doi.org/10.1016/j.accfor.2012.02.005

Miller, G. T., & Spoolman, S. (2007). *Environmental science: Problems, concepts, and solutions.* Toronto, ON: Brooks Cole.

Mohan, L., Chen, J., & Anderson, C. W. (2009). Developing a multi-year learning progression for carbon cycling in socio-ecological systems. *Journal of Research in Science Teaching, 46*(6), 675–698.

Myers, N. (2002). Environmental refugees: A growing phenomenon of the 21st century. *Philosophical Transactions of the Royal Society of London. Series B: Biological Sciences, 357*(1420), 609–613. https://doi.org/10.1098/rstb.2001.0953

National Council for the Social Studies. (1994). *National curriculum standards for social studies.* Silver Spring, MD: Author.

National Council for the Social Studies. (2010). *National curriculum standards for social studies: A framework for teaching, learning and assessment.* Silver Spring, MD: Author.

National Research Council. (2011). *A framework for K-12 science education: Practices, crosscutting concepts, and core ideas.* Washington, DC: The National Academies Press.

Ostrom, E., Burger, J., Field, C. B., Norgaard, R. B., & Policansky, D. (1999). Revisiting the commons: Local lessons, global challenges. *Science, 284*(5412), 278–282. https://doi.org/10.1126/science.284.5412.278

Peters, M. A. (2001). *Poststructuralism, Marxism, and neoliberalism: Between theory and politics.* Lanham, MD: Rowman & Littlefield.

Potter, E. F., & Rosser, S. V. (1992). Factors in life science textbooks that may deter girls' interest in science. *Journal of Research in Science Teaching, 29*(7), 669–686.

Rogers, R., Malancharuvil-Berkes, E., Mosley, M., Hui, D., & Joseph, G. O. G. (2005). Critical discourse analysis in education: A review of the literature. *Review of Educational Research, 75*(3), 365–416.

Rose, N. (1999). *Powers of freedom: Reframing political thought.* Cambridge, UK: Cambridge University Press.

Rudel, T. K., Defries, R., Asner, G. P., & Laurance, W. F. (2009). Changing drivers of deforestation and new opportunities for conservation. *Conservation Biology, 23*(6), 1396–1405. https://doi.org/10.1111/j.1523-1739.2009.01332.x

Rudolph, J. L. (2002). *Scientists in the classroom: The cold war reconstruction of American science education.* New York, NY: Palgrave.

Rudolph, J. L. (2003). Portraying epistemology: School science in historical context. *Science Education, 87,* 64–79.

Saad, L. (2011). Water issues worry Americans most, Global warming least. *Gallup Politics.* Retrieved from http://www.gallup.com/poll/146810/water-issues-worry-americans-global-warming-least.aspx

Schleppegrell, M. J. (2004). *The language of schooling: A functional linguistics perspective.* Mahwah, NJ: Erlbaum.

Schnaiberg, A. (1980). *The environment, from surplus to scarcity.* Oxford, UK: Oxford University Press.

Schnaiberg, A., Pellow, D. N., & Weinberg, A. (2000). *The treadmill of production and the environmental state.* Evanston, IL: Institute for Policy Research.

Sellers, C. C. (2012). *Crabgrass crucible: Suburban nature and the rise of environmentalism in twentieth-century America*. Chapel Hill, NC: University of North Carolina Press.

Sharma, A. (2012). Global climate change: What has science education got to do with it? *Science & Education, 21*(1), 33–53.

Sharma, A. (2013, April). Where are the people? understanding representations of society-nature relationships in a middle grades science classroom. *Paper accepted at the Annual Meeting of American Educational Research Association*, San Francisco, CA.

Sharma, A., & Anderson, C. (2009). Recontextualization of science from lab to school: Implications for science literacy. *Science & Education, 18*(9), 1253–1275.

Sharma, A. & Buxton, C. (2012, March). *Where are the people? Understanding representations of society-nature relationships in State Science Standards in United States*. Paper presented at the annual meeting of National Association for Research in Science Teaching, Indianapolis, IN.

Smolkin, L. B., McTigue, E. M., Donovan, C. A., & Coleman, J. M. (2009). Explanation in science trade books recommended for use with elementary students. *Science Education, 93*(4), 587–610.

Stake, R. (1995). *The art of case study research*. Thousand Oaks, CA: Sage.

Sterman, J., & Sweeney, L. (2007). Understanding public complacency about climate change: Adults' mental models of climate change violate conservation of matter. *Climatic Change, 80*(3), 213–238.

Takacs-Santa, A. (2007). Barriers to environmental concern. *Human Ecology Review, 14*(1), 26–38.

The New London Group. (1996). A pedagogy of multiliteracies: Designing social futures. *Harvard Educational Review, 66*(1), 60–92.

Tikka, P. M., Kuitunen, M. T., & Tynys, S. M. (2000). Effects of educational background on students' attitudes, activity levels, and knowledge concerning the environment. *The Journal of Environmental Education, 31*(3), 12–19. https://doi.org/10.1080/00958960009598640

Treanor, B. (2010). Environmentalism and public virtue. *Journal of Agricultural and Environmental Ethics, 23*(1), 9–28.

Tyson, H. (1997). *Overcoming structural barriers to good textbooks*. Washington, DC: National Education Goals Panel.

Weaver, A. A. (2002). Determinants of environmental attitudes. *International Journal of Sociology, 32*(1), 77.

Weiss, I. R., Pasley, J. D., Smith, P. S., Baniflower, E. R., & Heck, D. J. (2003). *Looking inside the classroom: A study of K–12 mathematics and science education in the United States*. Chapel Hill, NC: Horizon Research.

Westra, L. (2009). *Environmental justice and the rights of ecological refugees*. Sterling, VA: Earthscan.

Zak, K. M., & Munson, B. H. (2008). An exploratory study of elementary preservice teachers' understanding of ecology using concept maps. *Journal of Environmental Education, 39*(3), 32–46.

Zeitvogel, K. (2011). *50 million 'environmental refugees' by 2020, experts say.* Retrieved from http://www.google.com/hostednews/afp/article/ALeqM5j nW80NlFZ259UCgMAHSd3ekHutiQ?docId=CNG.aa651167cd0af745b-3cb395cf1d402e3.c41

CHAPTER 5

The Enacted Curriculum: Representations of Nature in Science Teaching

It was 7:30 in the morning, and the sun was out brightening the morning sky. After all, it was still August—the first month of the new school year at the Little Creek Middle School in a rural county in Georgia. As one of us (Sharma, and hereafter referred to in the first person) settled back in a chair in Ms. Rebecca Gilmour's seventh-grade life science classroom, a motley mix of sleepy and fresh-eyed faces trickled in and quietly found places to sit in the first few rows of student chairs. These were the volunteer members of the school Ecology Club. They had come 40 minutes before the start of school to attend the first meeting of the club. Facing the students were Ms. Gilmour and another teacher, Mr. Baker, who together ran this club as faculty sponsors. Soon there were 24 students in the room, and Ms. Gilmour opened the meeting by welcoming the students. She shared the purpose of the club—to promote environmental awareness through field trips and service projects—and then Mr. Baker presented a slide show that showcased the field trips and activities that members of this club had undertaken in previous years. After the slide show, both the teachers solicited suggestions from the students about places they could visit for their annual ecology field trip this year. A few students volunteered some tentative suggestions that were noted and acknowledged as possibilities by Ms. Gilmour. Soon the meeting was over, and the students went off to their first classes of the day.

As I closed my notebook, Ms. Gilmour, perhaps thinking that I may not have gotten good data for my research, came up to me and whispered,

© The Author(s) 2018
A. Sharma, C. Buxton, *The Natural World and Science Education in the United States*, https://doi.org/10.1007/978-3-319-76186-2_5

"we are definitely less preachy and more action in this club." But, she need not have worried. The meeting had indeed been very helpful to me in understanding the teachers' notions of the "environment" and the activities that they could do to increase students' awareness about it. Getting to know the "action" was indeed just as important as "talk" in my efforts to understand how the natural world was represented and understood in her science classroom. Besides, this was just the beginning of the academic year and I had a whole years' worth of science instruction ahead of me to observe in her classroom.

In this chapter, we present a case study of how one teacher represented nature while teaching seventh-grade science in her classroom. We have been unable to find comparable studies looking at how the natural world was represented in science classroom settings, but we do look to make connections with other (nonclassroom) research conducted both in the United States and abroad to extend our assertions to a more global level. Our claims regarding generalizability of this study's findings rest on the fact that we are able to situate our localized assertions in the wider theoretical space of discourse studies. Following Dyson and Genishi (2005), we believe that "the findings of any qualitative case study are not replicable, per se; they are a concrete instantiation of a theorized phenomenon" (pp. 6–7). That is, insights developed from an in-depth exploration of a particular case can help us understand a more abstract, global phenomenon in ways that the intended audience can easily recognize (Simons, 2009). Our study, as the readers will notice, is located in the broader theoretical space of societal discourses. The ability of dominant, global discourses to seep through local boundaries and influence local representations, practices, and institutions has been explored by many scholars, such as Bourdieu and Wacquant (2001), Fairclough (2003), Robertson (1995), and Rose (1999). Our study shows how such seepage happens in one specific case of the representation of the natural world and its relationship with the social world in a seventh-grade life science classroom.

Further, we are well aware that, unlike scientific research where one aims to generalize from the study of a representative sample to a population, the generalizability of our research, as in most qualitative case studies, works differently. It is our hope that once readers see how the theorized "universal" particularizes itself in one context, they will be able to transfer, adapt, and apply explanations emerging from that one study to their own and other contexts. In this way, our study aims at *naturalistic generalization*

(Stake, 1995). That is, "a form of generalization arrived at by recognizing similarities and differences to cases or situations with which readers are familiar" (Simons, 2009; pp. 164–165).

As in previous chapters, the analysis of classroom talk in this case study serves to highlight the ability of discourses to ontologically represent, as well as to reproduce, our material-social world for us. The classroom data was collected by one of us (Sharma) through ethnographic methods of participant observation of school science discourse during the academic year 2011–12 in Ms. Gilmour's seventh-grade science classroom at the Little Creek Middle School in a rural county in Georgia. I chose to study seventh-grade science discourse because it is in this grade that students in Georgia are introduced to fundamental ecology concepts in their science instruction, such as ecosystems, biodiversity, and matter and energy cycles, which are essential for an ecology-based understanding of the natural world and its relationship with humans. I (Sharma) am a middle-aged, cisgendered, Asian, nonnative male and a citizen of India. After living in the state of Georgia since 2008 and in the United States since 2001, I see myself as a hybrid and continually evolving composite of different cultural identities that may or may not neatly meet, match, or juxtapose with each other. Though Ms. Gilmour welcomed me to her classroom warmly, it took a while for the students to get used to my presence in the classroom. After a few days they learned to ignore me and allowed me to observe them and Ms. Gilmour like a fly on the wall. The class consisted of mostly white with a few African-American, Latin, and bi-racial 12–13-year-old students. Thus, owing to the obvious sociocultural and age differences, I could sense that most of the students remained reserved and a bit hesitant in speaking with me throughout the study. Ms. Gilmour, on the other hand, quickly promoted me as her confidant and while teaching frequently came up to my table to share her observations on students and thinking behind her pedagogical decisions.

In this study, I was able to supplement my classroom observations with loosely structured, open-ended focus group interviews with students and the science teacher Ms. Gilmour, and with review of relevant student artifacts and curriculum material. This data was then analyzed using complementary methodological approaches of critical discourse analysis and ethnography. Based on a constant comparative method, we followed a recursive process of open and closed coding and identification of themes (Emerson, Fretz, & Shaw, 1995; Fairclough & Wodak, 2004). Finally, we wrote this chapter using an analytically thematized narrative form to present the results emerging from our case study. We have used pseudonyms throughout to protect the confidentiality and anonymity of the research participants in this study.

We begin with a brief introduction of the community and the school where Ms. Gilmour taught, followed by a portrait of Ms. Gilmour and her students and a short description of how the course syllabus was organized for the entire year. The discourse in Ms. Gilmour's classroom was marked by stable, predictable patterns that complemented her approach to teaching science. We present these patterns, followed by a thematic analysis of how the natural world and its relationship with humans were represented in Ms. Gilmour's classroom. We end the chapter by drawing connections between the major themes of this classroom-level study and the analysis of science standards and textbooks presented in preceding chapters, in order to make some initial generalizations about school science discourse on nature in the United States.

SITUATING MS. GILMOUR WITHIN HER SCHOOL CONTEXT

As mentioned earlier, the Little Creek Middle School where Ms. Gilmour taught is situated in a small, rural county of Georgia with a population of around 32,000. It is one of the more prosperous and better-educated counties of Georgia with median household income ranging in the mid-seventy thousand dollars and 45% of the population having a bachelor's degree or higher. The community is predominantly white though of late there has been an increasing influx of minorities, especially Hispanics, and is thus rated as one of the fastest growing counties in the state. This region has recovered rather well from the great recession of 2008. The unemployment rate is low at 3.5% as a host of corporations have moved to this county and neighboring counties in recent years to take advantage of good geographical location, weather, low crime rates, and proximity to some well-regarded research universities. Politically speaking, the county leans strongly towards the Republican Party and can be called reliably conservative in terms of cultural and religious values. For instance, Ms. Gilmour once confided to me that she is particularly careful to not ruffle too many feathers whenever she teaches biological evolution to her students.

The Little Creek middle school is regarded as one of the higher performing schools in the state in terms of student scores on standardized tests. This is hardly surprising as the majority of students come from middle- or –upper-middle-class backgrounds. Since socially classed norms play a major role in constituting the dominant discourses on science and policy issues, a study of school science discourse set in such a school can help us better understand how broader societal and policy discourses on nature and society percolate through discursive boundaries to influence science instruction at a classroom level. In terms of infrastructure and facilities, the school does not compare well with the other newer middle school in the county. It is housed in an old building that could do with some repair and upgrading. However, academically, it could not be said that students fared any worse than at the other middle school. For instance, Ms. Gilmour's students had access to a well-equipped and much-used science lab adjacent to her classroom. There I saw students performing science experiments that could very well be the envy of most schools in the state of Georgia and beyond.

Ms. Gilmour was a white, middle-class and middle-aged female with a stern visage and strong presence that would have been daunting to her students had it not been often shattered by her ready smile and softened by her hard-to-be-missed concern for the well-being of her students. Ms. Gilmour was a veteran science teacher, as she had been teaching for 16 years all together and in this school for the last 13 years. Ms. Gilmour had two graduate degrees, one in plant pathology and one in secondary science education. After teaching a few years, Ms. Gilmour added middle grades and gifted education certifications to her academic accomplishments and just two years ago had also earned an education specialist degree in secondary science education. These academic credentials positioned her as a highly qualified teacher at her school, and she was spoken of highly by her colleagues. She was recently recognized as the "Teacher of the Year" by the school district and was also nominated for the Presidential Award for Excellence in Mathematics and Science Teaching.

Situating Ms. Gilmour Within Her Seventh-Grade Science Class

Ms. Gilmour started her career as an eighth-grade science teacher in a school in the neighboring county where she taught for three years. After shifting to the Little Creek Middle School, she initially taught Language Arts for one year and eighth-grade physical science for three years before transitioning to teaching seventh-grade life science. Thus, this was her ninth year teaching life science to seventh graders. Once while talking with her I asked her to reflect on how she liked teaching science to seventh graders. Ms. Gilmour replied

... while eighth graders are mature, seventh graders tend to do more stupid stuff ... But they are still interested in science. They are not cynical like when they get to high school. I think you have to have a sense of humor to teach middle school because they do stupid things ... that are kind of funny. I really enjoy it, and I like the life science part. The reason I got out of teaching eighth grade because that had become about physical science and I don't like physical science. I like life science ... I am a seventh-grade teacher who likes seventh-grade personalities better so it has worked out very well.

Thus, on most days when I entered Ms. Gilmour's classroom I found an engaged teacher looking comfortable and content teaching life science to a group of 12–13-year-olds. From what I could tell she cared deeply about her students and worked hard to teach science to the best of her abilities. In turn, the students were deferential in their interactions with her and readily gave her their attention whenever she demanded it. Ms. Gilmour taught several sections of seventh-grade science, but I focused my observations on her third period class based on her recommendation that third period would be the best to observe as it "represented a typical class of seventh graders." This section had 27 students of roughly even gender distribution. In terms of race and ethnicity, the class was fairly representative of the local community as white students vastly outnumbered minority students.

As we divulge in the next section, in many ways Ms. Gilmour came across as a science teacher cast in a traditional mold. This was well reflected in the way she preferred to physically organize her classroom space with rows of individual student desks neatly arranged to fill the room. A whiteboard and a projector screen were at the front along with a long table while Ms. Gilmour used a smaller office table in a corner at the back of a room whenever she needed a computer to work on. On one side of the classroom towards the back one could find some book shelves with various science magazines for students to read during free time. The walls were adorned with student projects; a few also hung down from the ceiling. Typical of many life science teachers, Ms. Gilmour kept an animal in the classroom: a small aquarium on a table towards the back of the room with a lonely looking snapping turtle emerging from the water to rest on a small platform now and then.

Ms. Gilmour and the School Science Discourse

I came to recognize a fairly predictable rhythm to Ms. Gilmour's teaching. During that academic year, Ms. Gilmour also had some student-teachers from a neighboring university in the classroom doing their science teaching apprenticeship. I found that they too more or less followed both the instructional rhythm and style of Ms. Gilmour's teaching. A typical lesson started with a warm-up activity that students found already displayed on the projector screen or the whiteboard as they entered the classroom. This lesson starter was usually in the form of questions that students had to answer. On most days, the warm-up questions were designed to check if the students remembered scientific vocabulary and definitions of scientific terms. Sometimes Ms. Gilmour had students read from their science textbooks for their warm-ups. This beginning routine was accompanied by announcements and instructions, such as about home assignments and upcoming tests. The warm-ups sometimes lasted for up to 20–25 minutes out of the 70-minute class period. This was because Ms. Gilmour thought it was important to review, correct, and elaborate upon student responses to warm-up questions. After the warm-ups were over, Ms. Gilmour led the students through the main instructional activities. These activities were almost always centered on a few "Essential Questions" that were aligned to the seventh-grade science curriculum as outlined in the state-mandated content standards, labeled as the Georgia Performance Standards (GPS). These questions were always listed on the white board in front of the classroom. The main instructional activities ranged from science experiments in the adjacent lab room to "read-alouds" from a science textbook. These activities lasted until a few minutes before the end of the class period, at which time students were given a few reminders regarding upcoming assignments, activities, and tests and were then dismissed.

We find Mortimer and Scott's (2003) analytical framework for characterizing the key features of the talk in school science classrooms very helpful in understanding the nature of school science discourse in Ms. Gilmour's classroom. The framework allows us to understand the communicative interactions between Ms. Gilmour and her students in terms of (a) *focus*, (b) *communicative approach*, and (c) *action*. The *focus* here refers to the teaching purposes and the content of the classroom interactions. An analysis of our classroom observations in terms of the *focus* of classroom interactions reveals that Ms. Gilmour's teaching was primarily oriented towards "introducing and developing the scientific story" and "guiding students to work with scientific ideas and supporting internalization" (Mortimer & Scott, 2003, p. 25). This instructional choice implied that Ms. Gilmour did not focus much on "opening up the problem" or "exploring and working on students' views" (p. 26). Thus, for instance, to teach energy flow through ecosystems, Ms. Gilmour first talked about the various roles that different organisms play in an ecosystem as producers, consumers, and decomposers, and then she presented the students with examples of producers, consumers, and decomposers in an ecosystem. That is, the focus of the instruction was very much on presenting the scientific story and then illustrating it with examples. Further, in close alignment with Ms. Gilmour's teaching purposes, we found that the content of the classroom interactions was dominated by a pattern that Mortimer and Scott (2003) identify as *description-explanation-generalization*. That is, much of the school science discourse was devoted to providing a scientific account of a phenomenon, entity, or system (*description*), using scientific concepts, ideas, and theories to explain phenomena recorded in curricular material or observed by students (*explanation*) and using specific examples to make generalized claims of a scientific nature (*generalization*).

Whatever the instructional activity might have been on a given day, one feature that almost always defined it was the high degree to which Ms. Gilmour controlled the nature, implementation, and outcomes of the instructional activities. The rare exceptions were the student research projects that she periodically had students do throughout the academic year. In all other instances, Ms. Gilmour decided upon the topic, the type of instructional activity, and its enactment. Students' preferences were neither sought by the teacher nor proffered by the students. Of course, students participated in the instructional activities, responded to her questions and comments, and sometimes even asked a few questions of their own accord. But student participation almost always hewed to the expectations that Ms. Gilmour had laid out for them at the start of class. Thus, based on Mortimer and Scott's framework, our analysis of Ms. Gilmour's communicative approach along two dimensions of *dialogic-authoritative* and *interactive-non-interactive* communication positioned her communicative approach as *interactive-authoritative.*

In terms of *communicative action* in the classroom, we saw *patterns of discourse* that were dominated by *initiation-response-evaluation* (IRE) type of triadic teacher-student exchanges or its variants, such as *initiation-response-feedback* (IRF) or *initiation-response-feedback-response-feedback* (IRFRF). That is, the classroom talk between Ms. Gilmour and students was almost always initiated by Ms. Gilmour and flowed back through her. This initiation was usually in the form of a question directed at a specific student, but sometimes also in terms of an open invitation for any student to participate. Students' responses were met with either an evaluative judgment or an elaboration that then moved the classroom discussion forward. Of course, sometimes students' curiosity or concern got the better of their reticence and led to student-initiated exchanges. For instance, one day while teaching about energy pyramids, Ms. Gilmour led an IRE type of exchange on ways organisms use energy they consume. This exchange led to numerous questions from students about eating habits of animals, the value of fat for survival, and how humans eat and get energy from food. Further, students sometimes initiated exchanges when they wanted clarifications on assignments and tests. IRE, IRF, and IRFRF types of classroom exchanges matched well with the nature of the *teacher interventions* that occurred in the classroom discourse. Analysis of classroom observations reveal that Ms. Gilmour's interventions were largely aimed at presenting the school science interpretation of the ways the biophysical world exists and works by: (a) *shaping* key scientific ideas and concepts by introducing scientific terms and differentiating between them, (b) *marking* them by reiterating them, and (c) *sharing* them by making them available to all students. Thereafter, Ms. Gilmour would devote her attention to (d) *reviewing* the scientific story so far presented and (e) *checking for student understanding* (Mortimer & Scott, 2003).

Ms. Gilmour was well aware that she needed to differentiate her instruction as not all of her students were on the same level in terms of their understanding of science and the ability to process unfamiliar scientific knowledge. For instance, one day she came up to my desk to share her worries about the fact that another differently abled student had joined her third period class the previous week, bringing the number of students in this class with extremely weak mathematics skills to five. In response to this diversity, I found that Ms. Gilmour often differentiated her quizzes and projects based on students' current class performance. But underlying this instructional differentiation, there always remained one constant— Ms. Gilmour's commitment to teach the normative school science interpretation of the world to all of her students. With this brief portrait of the overall contours of the school science discourse in Ms. Gilmour's classroom, we are now well placed to appreciate the way nature was represented through this discourse.

The Enacted Curriculum

Ms. Gilmour had divided the seventh-grade science curriculum into seven units, following the state standards and district guidelines. She began the year with a unit on processes that composed the *Scientific Method*. For the

Table 5.1 Unit sequence

Unit	Content	Months of the year
1	Scientific method	August
2	Environmental science	September–October
3	Cells and heredity	October–December
4	Organization of life	December–February
5	Genetics	March–April
6	Human body	April
7	Evolution	May

remainder of the academic year, she taught science content organized into six units of varying durations (see Table 5.1). When I asked her what she thought about the seventh-grade curriculum, she replied that "I think for the most part it's pretty good." Her one big complaint was that the state standards were not specific enough so it was difficult for her to know "… Exactly how in-depth are we supposed to go into some of these topics …" that she was supposed to teach. She also worried that the current state-mandated end-of-the-year standardized assessment, the "Criterion-Referenced Competency Test" (CRCT), did not require students to know very much about science content. As Ms. Gilmour opined, "I think a lot of the CRCT … it sounds terrible … CRCT in science for me is intelligence and reading ability … and our reading comprehension has increased substantially as a school. … So I think that's part of the reason why the test scores are up and … a lot of these tests are less picky about content and more about processes and … the kids can figure things out." The environmental science unit taught near the beginning of the academic year covered foundational concepts and ideas in ecology and environmental science, such as relationships between organisms in an ecosystem, energy flow and matter cycles in an ecosystem, adaptations of organisms to environmental conditions, characteristics of Earth's major biomes, and major environmental issues. The analysis presented below regarding representation of nature in Ms. Gilmour's classroom is primarily based on her teaching of this unit, as, not surprisingly, this unit is where the majority of explicit teaching about "nature" occurred. The representation of nature and its relationship with the humans in Ms. Gilmour's school science discourse was marked by three major themes, described in the following sections.

Nature as an Abstraction

It is said that enlightenment philosophers tended to see "nature as an abstraction or an inert object of study" (Caradonna, 2014; p. 52). Coming from a family of scientists, Ms. Gilmour had a similar, positivist standpoint on science when it came to teaching it to the students. Thus, nature or the natural world was presented to the students as a hierarchically organized network of interrelated concepts that referred to idealized representations of entities and processes found in the biophysical world. Such a conceptualization of nature was accomplished through three main instructional strategies. First, there was a strong emphasis on learning the abstract building blocks of the conceptual structure of nature *qua* abstraction through learning scientific vocabulary and definitions of scientific terms. These scientific terms familiarized students with the fundamental entities and processes that characterize the abstract representation of the natural world common in school science. While talking about her plans for the forthcoming year, Ms. Gilmour admitted, "The core thing as far as emphasis is … on vocabulary. A lot of that pretty much stays. So … I am pretty old fashioned at this thing." Thus, on most days warm-up questions centered on review of vocabulary and definitions. For instance, see Table 5.2 for some of the warm-up questions that students found displayed on the screen when they entered the classroom on 22 September.

Sometimes, Ms. Gilmour instructed students to take down notes from her PowerPoint presentations or from the science textbook. These notes mostly centered on writing down of definitions. Thus, assessment of students' recall of vocabulary and definitions was usually a big part of Ms. Gilmour's quizzes and tests. Table 5.3 presents a sample of questions asked in the first semester examination that year.

Second, learning about the natural world as an abstraction meant learning its conceptual structure, namely, the relationships between different abstract entities and processes that constituted the abstract ontology of the

Table 5.2 Warm-up questions

Identify the vocabulary term:
1. A living factor in an ecosystem.
2. A struggle between organisms over a limited resource.
3. A close relationship between different species and at least one species benefit.
4. Where one organism lives.
5. A symbiotic relationship in which both species benefit.

Table 5.3 Examination questions

First Semester Exam 2011-2012 D   Name _____
A   Date _____ Period _____

Matching - Place the letter of the vocabulary term to the left of the definition. (3 points each)

____ 1. A consumer that eats only plants.   A. carnivore

____ 2. A consumer that eats only animals   B. consumer

____ 3. A consumer that eats both plants and animals   C. decomposer

____ 4. The energy role of animals and other organisms that eat. This energy role is never at the base of an energy pyramid.   D. desert

____ 5. The energy role of organisms that make glucose (food). Plants and algae are examples.   E. grassland

____ 6. The energy role of organisms that break down dead organisms and waste. They rot things. Examples are bacteria and fungi.   F. herbivore

____ 7. A biome that receives less than 25 cm of precipitation per year. Organisms that live here are adapted to lack of water and extreme temperatures.   G. omnivore

____ 8. A biome that receives large amounts of rain and is located near the equator. It has greatest variety of species.   H. producer

____ 9. Trees in this biome lose their leaves during the winter. Temperatures vary during the year. There are seasons. Many animals hibernate in this biome. Oconee County is found in this biome.   I. temperate deciduous forest

____ 10. This biome has a lot of grass. There are many large herbivores like bison, deer, and antelope.   J. tropical rain forest

natural world. This is evident from the "Essential Questions" for the environmental science unit plan that Ms. Gilmour co-authored with the other two seventh-grade science teachers at her school (see Table 5.4). Here except for the second question about ecological issues, all other questions pertain to abstract entities, processes, and relations that build the conceptual foundations of ecosystem ecology. For instance, students in Ms. Gilmour's class learned that in terms of energy roles there are three types of organisms in an ecosystem: producers, consumers, and decomposers. Then they learned how producers, consumers, and decomposers are related

Table 5.4   Environmental science unit plan essential questions

LEARNING FOCUSED UNIT LESSON PLANNING FORM

Name: ▓▓▓▓▓▓▓▓▓▓▓▓▓▓▓▓   Class: 7th Life Science
Dates: 8/19/09-10/7/09
Unit 1: You Be The Ecologist

ESSENTIAL AND KEY QUESTIONS (KQ)	Unit Essential Questions: What is ecology? What are important ecological issues? How are energy and nutrients cycled through the ecosystem? What types of relationships exist between organisms? How do these relationships impact energy and nutrient cycling? How does energy move through and ecosystem? How are nutrients used and recycled in an ecosystem? What relationships are observed in ecosystems? How do these relationships influence energy and nutrient flow in an ecosystem? What are the aquatic and terrestrial biomes and the characteristics of each?

to each other through processes of matter cycles and energy flows. Similarly, students learned to hierarchically organize the natural world in terms of organisms, population, community, and ecosystem. This does not mean that Ms. Gilmour did not give examples to illustrate these abstract ideas and relations. She was always ready with examples for each topic she covered. But, it was also clear that in terms of curricular focus, abstract concepts were the key learning goals, while examples were simply there to support the understanding of concepts. Thus, the students in Ms. Gilmour's class learned to understand the world as an abstract system where biological, chemical, and physical processes connected different entities and exhibited certain well-defined emergent properties. As we saw in the third chapter, this was also a core defining perspective for conceptualization of ecology and environmental science standards in the Next Generation Science Standards as well as the Georgia Performance Standards.

Third, the conceptualization of the natural world as an abstract system meant that explanation was often reduced to explication (Braaten & Windschitl, 2011). That is, Ms. Gilmour used her explanations to unpack the dense school science discourse so as to make it accessible to her students. "Explanation as explication" can be seen in the following instructional episode that happened when Ms. Gilmour was introducing the topic of *food chain* to the students. After a warm-up in which students classified various organisms in terms of the energy roles (producers, consumers, or decomposers) they played in an ecosystem, Ms. Gilmour showed them a PowerPoint slide with the following definition of a *food chain*: "Food Chain = A series of events in which one organism eats another." Ms. Gilmour asked students to copy it down in their notebooks. Then she went to the whiteboard and wrote down the following food chain:

$$\text{Grass} \to \text{mouse} \to \text{snake} \to \text{hawk}$$

With the help of students, Ms. Gilmour then labeled each organism in this food chain according to their energy role. That is, grass, mouse, snake, and hawk were labeled as producer, consumer (first level), consumer (second level), and consumer (third level) respectively. Then Ms. Gilmour further explained:

> I call the food chain an eating chain because we don't have any decomposers here. All we see here is eating ... that's what I call it. This is not something you are going to see anywhere else. So you have your producers, and you have your consumers. ... In a correctly done food chain, there are no decomposers.
> ... But you may see incorrectly done food chains. You may see decomposers and you may also see Sun in the beginning. ... But that is really not correct.

As we can see here, Ms. Gilmour's explanation is clearly aimed at further clarifying the meaning of *food chain*. Sometimes, we saw "explanation as explication" in terms of elaboration of or sharing of scientific reasoning behind a phenomenon. For instance, in the following exchange between a student and Ms. Gilmour:

Megan: Why are Venus flytrap plants called producers when they eat insects?

Ms. Gilmour: Venus flytrap and other plants that eat insects live in areas where there is poor soil. They are producers because they are making food ... they are green. They take sunlight and they make food. But they need other nutrients to be healthy ... like vitamins. You take vitamins to be healthier. There are certain things that are not in their soil ... the soil is so poor ... so they need vitamins. And they get them from the animals they eat.

Here we can see that Megan had asked a very good question that warranted a scientific explanation. In response, therefore, Ms. Gilmour explained the scientific position by detailing the reasoning behind Venus flytrap's apparently contradictory classification as a producer.

In addition, we also saw examples of "Covering Law" scientific explanations in which particular events and phenomena were represented as reasonable, predictable consequences of law-like regularities that the natural world as a system can be expected to exhibit (Braaten & Windschitl, 2011). For instance, one day a student-teacher from a neighboring university, Ms. Smith, showed video clips of the movie "Finding Nemo" to teach a lesson on the interactions between organisms in an ecosystem. In the middle of this lesson, Ms. Gilmour took over the instruction from Ms. Smith and pointed the students towards the clownfish and sea anemone relationship in the movie as an example of *mutualism*—a relationship between members of different species in which both organisms benefit. Another day Ms. Gilmour asked students to read aloud a section on *Ecosystem* from the science textbook and take down notes on the important scientific ideas and terms from the section. Then she showed the class a YouTube video of a forest ecosystem to point out the different *biotic* and *abiotic* factors in that ecosystem. In such instances, we see Ms. Gilmour using the "Covering Law" model of scientific explanation a little differently by first teaching students about patterns found in the natural world and then illustrating them through specific examples.

The advantages of representing the natural world as an abstract system in scientific thinking cannot be denied. By doing so, both scientists and science teachers are able to temporally arrest transitory events in the world to study them, to assign fixed properties and organize them into classificatory schemes and stable phenomena that can be studied and more easily represented. But this affordance comes at a cost: the representations of the world begin to appear stable and transcendent and thus somewhat distant and different from the world they claim to represent. How that happened in Ms. Gilmour's classroom is what we turn to next.

Nature as a Stable World

It was late December, and the school was just a couple of days short of closing for the winter break. The environmental science unit had been completed, and the class had moved on to the next unit on *Cells and Heredity*. But Ms. Gilmour thought that doing a review of the content covered in the entire environmental science unit was in order. So, as I entered her classroom that day, I found students sitting in groups of four at tables that had been arranged for a quiz show type of game. The class was all set to play what Ms. Gilmour called the "review game" on the environmental science unit. She would ask a question to a group. If the group answered it correctly, it received points. If they could not answer or offered an incorrect answer, the question went to the next group. At the start of the game a few students jestingly made a few faux complaints on the composition of different teams, but on the whole, the students played the game enthusiastically. They also did quite well answering Ms. Gilmour's questions that are listed in Table 5.5. These questions can be analyzed in many ways, but for now, we will focus on what these questions reveal about the temporal aspects of the world they describe. Quite clearly, time is conspicuously absent in these questions, which speak of a world that doesn't change with the passage of time. Of course, the questions on their own offer only a partial and blurred view of the school science discourse. But as we analyzed other pieces of evidence from this unit, these review questions seemed to fit rather nicely with the emerging theme of the school science discourse in Ms. Gilmour's classroom that represents a world where time has ceased to matter in the overall representation. Let us, therefore, consider a few other examples from the school science discourse that supports this theme.

Understanding the natural world in terms of a hierarchically organized system of individuals, populations, communities, and ecosystems was a

Table 5.5 Questions in the review game

1. What is the energy role in an ecosystem?
2. What is the difference between biotic and abiotic factors?
3. What is the original source of energy for an ecosystem?
4. What is the energy role at the base of an energy pyramid?
5. What is the percentage of energy loss from one trophic level to another?
6. Give an example of parasitism.
7. Give an example of mutualism.
8. Give an example of commensalism.
9. What is the definition of symbiosis?
10. I will describe a biome. You will tell me which biome it is.

major topic of the environmental science unit. In one of the activities designed to teach this idea, Ms. Gilmour asked students to open the science textbook to pages 20–21. These pages show a pictorial representation of ecological organization in the prairie ecosystem. Then based on this representation, Ms. Gilmour asked students to draw a similar graphic representation of an ecosystem in Africa. The classroom interaction went something like this:

Ms. Gilmour:	Okay, first thing you are going to do. ... Divide the page into four sections. The left is organism ... Next is what. ... A group of similar organisms?
Jessica:	Populations.
Ms. Gilmour:	Yes, populations. ... And the community. ... And the last one is ...
Elizabeth:	Ecosystem.
Ms. Gilmour:	Ecosystem ... We are going to do African Savannah. There are lots of animals there. You saw in the African Wildlife video ... So over in this right picture (pointing to her own drawing of the African Savannah that she was making on the whiteboard) ... My drawings are not the greatest. So if I can put my drawings on the board, you can put your drawings too. So what am I going to want in my African thing? ... I guess I am going to want a ...?
Zack:	A lion.
Ms. Gilmour:	Yes. A lion ... You are going to want to draw an ecosystem that is taking place in Africa ... I will have just one lion because you want just one (unclear) the predator. I want to have elephants, zebras, and giraffes .... So start putting in things ... You want to make it seem like an African savannah. So ... let's see .... I want an elephant.
Zack:	Elephant is my favorite animal.
Ms. Gilmour:	Mine too. ... So quietly do this in pencil. And if I can put my drawings on the board, surely you can put yours on paper because my drawings are worse than yours. ... Quietly do this. ...

While drawing her representation on the whiteboard, Ms. Gilmour continued:

Ms. Gilmour:	Do your own drawings and let me do mine. ... There is my elephant ... My drawings are terrible ... Okay, I need to have trees. ... Skimpy trees. You need some plants in the ecosystem. You need to have more than one of each kind. ... So I am going to have more giraffes there.
Zack:	I am going to draw elephants.
Peter:	I am putting in an Ostrich!
Jessica:	Can I have some water in mine?
Ms. Gilmour:	Yes, you can. But do your own, this is just mine. I am just trying to give you something that you can ... But you have got to do your own. You have to have several animals. ... I should put some birds up there ...
John:	Do turtles count?
Ms. Gilmour:	Turtles, yes. ... Okay, I am putting the watering hole in here. ... You got the colored pencils, right? If you got the colored pencils, get them out. ... and then I need to have my clouds. ... Make sure you have more than one example for each species ... You can't have just one so that you can show a population. Okay, you got to have plants you know. ... Now, don't forget that in your ecosystem you need to have abiotic factors. So that is why I have my watering hole up there ... I will put some leaves on my trees. ... So you will have to pick one of your organisms that you have more than one ... like I have a bunch of giraffes ... So I am going to shove a giraffe over here (puts a drawing of giraffe in the column for the organism) as an organism. You have to have one individual in here.

The science textbook used by Ms. Gilmour in this activity does not mention any temporal changes when it presents the ecological organization of life in terms of native flora and fauna. Neither did the National Geographic video clip on the African savannah that Ms. Gilmour showed to the class a few days prior to this activity. Thus, it is likely that ecological changes are not something that students will think of when asked to graphically represent an ecosystem. As can be seen in this exchange, Ms. Gilmour too, in her teaching, perpetuated the notion that the natural systems, such as African Savannah, are stable systems that continue to exist as they have since times immemorial. This notion stands in strong contrast to current understanding among scientists that life on the African Savannah is anything but stable in the twenty-first century. According to current description, "Large areas of the savannah are under stress and disturbance from human activities such as grazing, fuel wood, and timber collection and land clearing for cultivation" (Kalipeni, 2007; p. 1565). In fact, there are now strong reasons to fear that because of increasing atmospheric concentrations of CO_2, large areas of African Savannah may turn into woody forests by the end of this century (Higgins & Scheiter, 2012). However, during the entire class period when the ecosystem drawing activity was done, and in the preceding as well as succeeding class periods, Ms. Gilmour never once raised the issue of temporal changes in ecosystems on account of local or global factors. Similarly, in the student-teacher's lesson using clips from the movie "Finding Nemo," mentioned earlier, neither Ms. Smith, the student-teacher, nor Ms. Gilmour made any mention of the profound and ongoing ecological changes to the marine ecosystems shown in the movie.

It is not that ecological changes were completely omitted from the enacted curriculum. For example, Ms. Gilmour had students complete a research project in which they each picked one ecological challenge to research on their own and then present to the class. But ecological disturbances were never represented as causes that fundamentally or significantly changed the natural world. Ms. Gilmour was a highly knowledgeable science teacher, and she was clearly aware that ecological change is a constant feature in the natural world. For instance, one day a student, Jessica, asked her why their region had so many pine trees. Ms. Gilmour first lobbied the question back to the students. When none of the students could respond, Ms. Gilmour gave a detailed explanation of how the original forests in this region had been dominated by oak trees that were cut by settlers in the eighteenth and nineteenth centuries to clear agricultural land for growing cotton. As cotton farming gradually declined in the region in the late nineteenth and early twentieth centuries, the fallow land was planted with pine trees as a good early ecological succession tree species for this region.

So, when we present evidence that indicates that Ms. Gilmour portrayed the natural world as a stable world, the point is not to critique Ms. Gilmour's own knowledge base or teaching approach. Rather, these episodes should be seen as emblematic of school science discourse in a classroom where the science teacher consciously decides to teach science in ways that are expected of her in accordance with the state science standards. As presented in the third chapter, the science curriculum as mandated by the state science standards had a clear stance when it came to positioning humans in relationship with the natural world. As we present in the next section, Ms. Gilmour maintained a fidelity to the official school science view of the place of humans on the planet.

Nature as a Human-Free World

Nature videos were an important curricular material in Ms. Gilmour's classroom. One day she decided to show her students the National Geographic video "Amazon: Land of the Flooded Forest" to help the class understand how different organisms have developed adaptations to survive in their habitat. She wanted the students to watch the video and make a list of organisms along with their adaptations. Before starting the video, as she was giving instructions for the activity, she had the following exchange with a student:

Ms. Gilmour: We are looking at it (the video) from a biological standpoint. We are looking at different organisms shown in the video and their unique adaptations. The beginning part of this video ... this is an old National Geographic video ... and it is about the formation of the Amazon River and plate tectonics ... and then there is something about the people who live in this area, and also the organisms that are there. ... So you are not taking notes while you are watching. I will stop if you want to jot something. But I will be stopping periodically and you will be listing organisms and some of its adaptations ... As you watch focus on these unique animals and plants and how they are able to survive ... and ... yes?
Samantha: Will we add people as organisms in filling the chart?
Ms. Gilmour: No, we are not adding people as organisms ... although they are organisms ... and that's a great question ... We are focusing on the ... "wilder" aspects of this.

Indeed, in almost all video and pictorial representations of the natural world that Ms. Gilmour showed in her classroom, human beings or any signs of their influence were conspicuously absent. She made frequent use of National Geographic and Planet Earth videos in her teaching. These videos tended to show Earth as a human-free planet. Further, as we saw in the above exchange even when such representations did show people as part of the natural world, Ms. Gilmour instructed students to ignore them, claiming that studying about nature in her class meant that students should only focus on the "wilder" aspects of our world. This was true even when students were asked to use their own community to find examples of ecological concepts. For instance, after teaching about food webs for three days, Ms. Gilmour asked students to make a food web for their local county as a homework assignment, but asked them to leave out all human inhabited areas as she said, "We are talking only about the natural areas" of the county. We know, however, that food webs all over the world have been severely impacted by human activity, leading to the large-scale trophic downgrading of the entire planet (Estes et al., 2011; Strong & Frank, 2010). In light of such sobering scientific analyses, positioning of suburban areas as spaces where "natural" food webs can be found and analyzed might be seen as a discursive construction of the natural world as a space from which all humans and their influences have been erased. We saw this discursive erasure of humans from the field of study for ecological phenomena as a prominent leitmotif in Ms. Gilmour's teaching. When Ms. Gilmour began teaching the ecology unit, she had students copy the following definition of ecology from one of the science textbooks used in the classroom: "The study of how living things interact with each other and with their environment is called **ecology** (emphasis in original)" (Padilla, Cyr, & Miaoulis, 2002; p. 20). As we saw in her teaching of this unit, however, human beings were excluded from the category of living things for the study of ecology. Thus, following the state science standards' and science textbook's stance on the world, the school science discourse in Ms. Gilmour's classroom discursively partitioned the Earth into two independent and distinct, albeit interacting, parts—a natural world without humans and a social world of humans.

Externalization of the "social" from the "natural" meant that humans were overwhelmingly positioned in two complementary roles vis-à-vis their relationship with nature. These roles correlated with the type of topic under consideration. Indeed, the topics covered throughout the environmental science unit could be divided into two overlapping disciplinary areas—ecology and environmental science. In topics that related more closely to ecology, we found that the humans were overwhelmingly absent, or else were interpolated in the school science discourse as scientists who studied nature to uncover underlying ecological processes and to illustrate theorized phenomena through scientific evidence. Ms. Gilmour frequently talked about how ecologists study the natural world. Thus, the natural world was represented as a "world out there" that ecologists visited for the purpose of studying it. The environmental science unit plan was given the name "You be the ecologist." Ms. Gilmour sought to translate this pedagogical intention by having several inquiry activities for students in which they had to take on the role of ecologists to investigate and complete the activity. For instance, Ms. Gilmour bridged the first instructional unit on "Scientific Method" with the second "environmental science" unit with an inquiry project in which students acted as citizen scientists to explore an interesting and important natural phenomenon related to monarch butterflies. In this project, in collaboration with the local university and the nationwide research network "Journey North" (https://www.learner.org/jnorth/monarch/) students grew monarch butterflies in their classroom and studied their life cycle. Then they tagged and released them in time to let them migrate with other monarch butterflies on their annual migration journey to Mexico.

On the other hand, in topics that related to environmental problems, with a stronger affinity to the disciplinary area of environmental science, humans came across as the external harmful influence on natural systems. Interestingly, this harmful impact was largely presented in the contexts of pollution of natural resources and not in cases where ecological damage is caused by unsustainable resource extraction. Thus, for instance, on the topic of aquatic biomes and in reference to the essential question: "How do humans affect an aquatic ecosystem?" Ms. Gilmour showed the students the documentary "Chattahoochee: To fall in love with a river" that focused on pollution in the Chattahoochee River caused by disposal of untreated water by cities through which the river flows. Ms. Gilmour also made good use of topical issues to illustrate how humans impact ecosystems. For instance, I witnessed a lesson on how the Deepwater Horizon oil spill impacted the surrounding aquatic ecosystems. However, in these cases the key issues were usually the ecological concepts and not the environmental damage. The topical issues were used rather as a hook to engage the students so that standards-based concepts could be taught. When I asked Ms. Gilmour how she integrated science-society issues in her teaching, she responded:

> I am not consciously putting it there. ... But like the oil spill ... that really ties in with a variety of things we do ... you have an abiotic factor that is affecting things so you want to talk about the objectives in environmental science ... abiotic factors versus the biotic factors. It is also affecting the survival of species which get to the other objectives that have to do with genetics, such as breeding and evolution ... that stuff. So that fits in.

All of us are differently positioned when it comes to our relationship with the nonhuman biophysical world. In this relationship spectrum one end may be represented by forest-dwelling indigenous communities that have a direct metabolic relationship with their immediate biophysical environment, while on the other extreme end we find denizens of advanced capitalist societies that pollute and consume resources from ecosystems all over the world. But in alignment with the state science standards, the school science discourse in Ms. Gilmour's classroom did not differentiate between people in this way. Everyone was seen as equally culpable as it was "humans" or "people" who were represented as polluting or harming the natural world. At times, some specific group got named as the primary culprit, such as the city of Atlanta in the case of polluting of Chattahoochee River or the British Petroleum Corporation in the case of Deepwater Horizon oil spill. But the important thing to notice in such instances is that even when specific actors were named, their action and role was not presented as an illustration of the differential role that different socioeconomic collectives play in creating and aggravating ecological crises. Thus, opportunities for a more illuminating socioeconomic critique of human-nature relationships were missed perhaps in favor of better compliance with the state-mandated curricular expectations.

Conclusion

Over the years, Ms. Gilmour had become increasingly frustrated by the deprofessionalization of teaching that she attributed to a reduction in teacher autonomy and an increase in accountability measures. She also hated the idea that the state may soon be basing her salary on her students' performance on state assessments. But, as far as teaching science was concerned, it was rare to hear her complain. She liked the science content she taught, as it was represented in the state standards and textbooks. She only wished that the state would be "more specific in their standards" so that teachers knew "... Exactly how in-depth are we supposed to go into some of these topics." Having received her own science education in the field of plant pathology many years back, she was seeped in traditional science discourse that positioned scientists and official sources of knowledge, such as science textbooks, as presenting authoritative accounts of what the world is like. Thus, when we observe her teaching science using an interactive-authoritative communicative approach and guiding initiation-response-evaluation type of teacher-student exchanges, it is hard not to see a mutually reinforcing complementarity between her belief in mainstream, official accounts of science and her teacher-centered pedagogies. Ms. Gilmour believed in the mainstream authoritative scientific account of the natural world and worked hard to authoritatively disseminate it among her students. Not surprisingly, then, she did not tap local funds of knowledge or everyday discourses that resided in the community of her students, nor did she seek to integrate nonmainstream or indigenous perspectives on nature into her science teaching. In this regard, as with other aspects of her teaching that we have discussed, we do not wish to claim that Ms. Gilmour is different from the vast majority of her science teaching peers. As Lee and Buxton (2010) found in their review of the relevant literature, while there is evidence that pedagogical approaches grounded in students' cultural backgrounds and everyday knowledge can make a difference in their science learning, a combination of education policies, standards, and restricted perspectives on learning has resulted in limited efforts to prepare science teachers in the use of such approaches. Indeed, examples of science teachers integrating culturally relevant pedagogies and community funds of knowledge into their classroom practices in the United States (e.g., Buxton, 2006; Barton et al., 2008; Warren, Ballenger, Ogonowski, Rosebery, & Hudicourt-Barnes, 2001) stand out as stark exceptions to the rule. Thus, Ms. Gilmour was not much different from most science teachers when she eschewed dialogizing school science discourse with local funds of knowledge, everyday discourses, and traditional ecological knowledge.

In the third chapter we discussed how a Cartesian-Newtonian mechanistic perspective on the world influences the way our world is presented in national and state science standards. In this view, the natural world is a physical reality external to and different from our social world that can be objectively observed and understood using scientific theories and methods. As presented in this chapter, Ms. Gilmour worked to share this perspective with her students. In her classroom, the world was represented through a dualism that separated the natural from the social with a clear understanding that the purview of science is restricted to the discursively created "natural world" alone, where "wilder" interactions dominate. Further, the school science discourse in her classroom represented the natural world as an abstract system of conceptually dense terms that were linked together in a hierarchically organized network of relationships. Though Ms. Gilmour tried to use illustrative examples of abstract ecological ideas and processes, the abstract ontology of nature appeared removed from students' lived experiences in the world. Not surprisingly, therefore, when Ms. Gilmour asked students to pick an environmental issue to research and present to the class most students picked an issue far removed from their own lived experiences.

Traditional science discourse still hews to a belief in the "balance of nature" metaphor that portrays a stable world that exists in a state of equilibrium marked by harmony and balance (Jelinski, 2010). Ms. Gilmour did not broach the issue of disturbed ecosystems trying to revert to their earlier ecological equilibrium. But her representation of the natural world as a stable world where ecological processes and phenomena exist independent of time certainly supported the "balance of nature" view that students may have or are likely to encounter in media, everyday discourses, and future science classes. The "balance of view" metaphor shares a close affinity with American culture and mainstream environmental values (Allchin, 2014; Simberloff, 2014). Besides, stable systems are easier to comprehend and teach. Thus, even if Ms. Gilmour had tried to present nature as dynamic, chaotic, and continually buffeted by disturbances, chances are that she would have faced strong headwinds in her efforts, especially considering that students in her classroom hailed from a community that had not faced noticeable ecological changes in the living memory of the students. Reduction of complexity was also achieved by externalizing humans from the natural world. This simplification may have made science content easier for students to comprehend. However, simple representations of our world also lay the foundations for an incorrect and ecologically harmful lifelong orientation towards the world and our place in it. Further, we do not know of any persuasive argument that establishes the futility or marginal value of exposing students to a scientifically up-to-date view of the world.

Scientific discourse comes across as highly authoritative to participants who are marginal to its production. Science teachers belong to this category of peripheral members in the wider scientific discursive community even though they play a vital role in disseminating this discourse in the society. Science discourse, especially in school settings, also tends to be a hegemonic discourse as it actively delegitimizes discourses deemed unscientific and normalizes certain perspectives and representations of our world for both teachers and students. Additionally, reduced teacher autonomy and heightened teacher accountability have further reduced incentives for teachers to teach anything other than the officially authorized ontology of the world. Thus, it is hardly surprising that Ms. Gilmour taught in the way we have outlined in this chapter. Perhaps, if we were in her shoes, we would not have acted much differently.

If we truly wish to see a different school science discourse in our classrooms, we need to attend to the wider network of relations in which science teachers find themselves located once they enter those classrooms. But to do so will require a more comprehensive and rigorous account of nature in school science discourse than what we could present based on our study alone. Efforts to help science teachers represent the world to students in ways that align with current understandings among scientists must be accompanied by a research agenda that addresses the acute paucity of research on this issue.

References

Allchin, D. (2014). Out of balance. *The American Biology Teacher, 76*(4), 286–290. https://doi.org/10.1525/abt.2014.76.4.13

Barton, A. C., Tan, E., & Rivet, A. (2008). Creating hybrid spaces for engaging school science among urban middle school girls. *American Educational Research Journal, 45*(1), 68.

Bourdieu, P., & Wacquant, L. (2001). New liberal speak: Notes on the new planetary vulgate. *Radical Philosophy*, (105). Retrieved from http://www.radicalphilosophy.com/default.asp?channel_id=2187&editorial_id=9956.

Braaten, M., & Windschitl, M. (2011). Working toward a stronger conceptualization of scientific explanation for science education. *Science Education, 95*(4), 639–669.

Buxton, C. A. (2006). Creating contextually authentic science in a "low-performing" urban elementary school. *Journal of Research in Science Teaching, 43*(7), 695–721.

Caradonna, J. L. (2014). *Sustainability: A history.* Oxford, UK: Oxford University Press.

Dyson, A. H., & Genishi, C. (2005). *On the case.* New York, NY: Teachers College Press.

Emerson, R. M., Fretz, R. I., & Shaw, L. L. (1995). *Writing ethnographic fieldnotes.* Chicago, IL: The University of Chicago Press.

Estes, J. A., Terborgh, J., Brashares, J. S., Power, M. E., Berger, J., Bond, W. J., ... Jackson, J. B. (2011). Trophic downgrading of planet earth. *Science, 333*(6040), 301–306.

Fairclough, N. (2003). *Analysing discourse: Text analysis for social research.* London, UK: Routledge.

Fairclough, N., & Wodak, R. (2004). Critical discourse analysis. In T. A. van Dijk (Ed.), *Discourse as social interaction* (pp. 258–284). Thousands Oak, CA: Sage.

Higgins, S. I., & Scheiter, S. (2012). Atmospheric CO_2 forces abrupt vegetation shifts locally, but not globally. *Nature, 488*(7410), 209–212.

Jelinski, D. E. (2010). On the notions of mother nature and the balance of nature and their implications for conservation. In D. G. Bates & J. Tucker (Eds.), *Human ecology: Contemporary research and practice* (pp. 37–50). Boston, MA: Springer US.

Kalipeni, E. (2007). Savanna (or tropical grassland). In P. Robinns (Ed.), *Encyclopedia of environment and society* (pp. 1564–1565). Thousand Oaks, CA: Sage Publications.

Lee, O., & Buxton, C. A. (2010). *Diversity and equity in science education: Research, policy, and practice*. New York, NY: Teachers College Press.

Mortimer, E. F., & Scott, P. H. (2003). *Meaning making in secondary science classrooms*. Philadelphia, PA: Open University Press.

Padilla, M. J., Cyr, M., & Miaoulis, I. (2002). *Science explorer: Environmental science*. Upper Saddle River, NJ: Prentice Hall.

Robertson, R. (1995). Glocalization: Time–space and homogeneity–heterogeneity. In M. Featherstone, S. M. Lash, & R. Robertson (Eds.), *Global modernities* (pp. 25–44). London, UK: Sage.

Rose, N. (1999). *Powers of freedom: Reframing political thought*. Cambridge, UK: Cambridge University Press.

Simberloff, D. (2014). The "balance of nature" – evolution of a Panchreston. *PLOS Biology*. https://doi.org/10.1371/journal.pbio.1001963

Simons, H. (2009). *Case study research in practice*. London, UK: Sage.

Stake, R. (1995). *The art of case study research*. Thousand Oaks, CA: Sage.

Strong, D. R., & Frank, K. T. (2010). Human involvement in food webs. *Annual Review of Environment and Resources, 35*, 1–23.

Warren, B., Ballenger, C., Ogonowski, M., Rosebery, A. S., & Hudicourt-Barnes, J. (2001). Rethinking diversity in learning science: The logic of everyday sense-making. *Journal of Research in Science Teaching, 38*(5), 529–552.

CHAPTER 6

The Received Curriculum: Nature as Understood by Students

Zach, Peter, Jessica, Megan, and one of us (Sharma[1]) were huddled around a small table in the lobby of the Little Creek Middle School one morning in the spring of 2012. Ms. Gilmour had let them out of the classroom so that I could talk to them. I had been with Ms. Gilmour's students observing them learn science for many months. The students had gotten used to seeing me in the classroom, and a degree of mutual trust and conviviality had developed between us. I felt that it was now a good time to talk to them about their views on learning about ecology and environmental issues. After listening to their ideas about environmental issues and what we should do about them, I remarked, "These are all great ideas ... where did you learn all that?" Zack smiled and responded, "I really form my own opinions on what should be done." I probed again, "So what is your source of information? The classroom?" To my surprise, Zack shook his head and muttered, "TV shows ... they talk about it all the time."

In the last chapter we talked about how, as a conscientious science teacher, Ms. Gilmour worked hard to teach science in her classroom. The students too, for the most part, appeared to pay attention in the classroom and participated in classroom learning activities as expected by the teacher. However, as I talked to her students it dawned upon me that there were far more powerful factors that shaped their views of the nature than Ms. Gilmour's teaching. Later as I read about what the existing research had to say on this topic, it became clear that these students illustrated rather well what other researchers had also been reporting about students' views

© The Author(s) 2018
A. Sharma, C. Buxton, *The Natural World and Science Education in the United States*, https://doi.org/10.1007/978-3-319-76186-2_6

in the United States as well as abroad. That is, though schools continue to be an important formative influence on students' views, other factors such as mass media exert far more influence on how students come to perceive nature (Kellert, 2002; Özdem, Dal, Öztürk, Sönmez, & Alper, 2014; Rickinson, 2001). With the rise of Web 2.0 the influence of an increasingly diverse mass media landscape has only grown multifold in the last two decades as "mass-mediated representations of nature that now appear online continue to reflect and inform how people think about the natural world" (Elliot, 2016; p. iii).

Thus, shaped by powerful influences over which they have limited control, students acquire orientations towards and understandings of the natural world that, as we show in this chapter, are not that different from those exhibited by adults in the United States and other advanced capitalist societies. Beginning with students' understanding of major ecological topics covered in science curricula, we focus on how students perceive nature and how they reason about environmental issues that they are familiar with. The students' views and understanding presented in this chapter are shaped both by existing research on the issue and by what Ms. Gilmour's students told me (Sharma) during my ethnographic study of her teaching which we explored in the previous chapter.

Students' Understanding of the Natural World

Research is not very optimistic about how well students understand their biophysical world from a scientific perspective. It shows that, on the whole, students "display considerable confusion about the science of environmental issues, often characterized by persistent misconceptions" (Rickinson, 2001; p. 232). For instance, students often can't distinguish between different phenomena, such as climate change and ozone depletion (Allen, 2014). They also exhibit considerable lack of understanding regarding processes and mechanisms underlying ecological phenomena. For example, a study on students' understanding of ecosystem concepts reported that students often hold naïve conceptions regarding cycling of matter in ecosystems (Jordan, Gray, Demeter, Lui, & Hmelo-Silver, 2009). That is, on the whole, it appears that despite a strong focus in school science on understanding the biophysical world from an ecosystem ecology perspective, most school students struggle with understanding our planet as a biogeochemical system (Eilam, 2012). In my conversations with students, I too was struck with the fact that after a few months of instruction by

Ms. Gilmour, most students could not recall much of the science content they had learned in the ecology unit. When pushed, the most they could do was to recall the nature videos they watched in class, field trips they took during the unit, names of a few ecological biomes, and a few scientific terms, such as food chain and food web.

Further, research indicates that students tend not to use whatever little scientific knowledge they do possess when making decisions in citizens' roles (Covitt, Tan, Tsurusaki, & Anderson, 2009). In fact, in a study with ninth-grade students, Cobern (2000) found not only that students did not integrate their scientific knowledge with their everyday thinking while engaging in discussions on environmental issues, but also that their success as students was poorly correlated with their use of science to make sense of their everyday world. Of course, these results should not lead us to conclude that students do not have distinct views and attitudes about the natural world and their relationship with it. In addition to school science, they live, perform, and are in fact constituted by powerful out-of-school discourses that shape their practices and perceptions of the world around them.

Before we begin outlining students' perceptions of nature, it needs to be acknowledged that students seem to treat "nature" as synonymous with "environment" (Payne, Cutter-Mackenzie, Gough, Gough, & Whitehouse, 2014). I found that Ms. Gilmour's students too used these two terms interchangeably. Research also shows that students hold diverse conceptions of nature. In an extensive study, Loughland, Reid, and Petocz (2002) asked over 2000 Australian students to complete the following sentence "I think the term/word environment means ...." Analysis of students' responses revealed six distinct conceptions of the environment that ranged from "from the least sophisticated—environment as a place—to the most inclusive and expansive—environment and people in a relationship of mutual sustainability" (p. 187). These results correspond well with another study that explored the mental models that students in US schools had about the environment (Shepardson, Wee, Priddy, & Harbor, 2007). These researchers found that the students deployed four kinds of mental models of the environment to complete an idea-eliciting "environments task." These models were: "Model 1, the environment as a place where animals/plants live—a natural place; Model 2, the environment as a place that supports life; Model 3, the environment as a place impacted or modified by human activity; and Model 4, the environment as a place where animals, plants, and humans live" (p. 327). Students displayed a distinct preference for Model 1. Not surprisingly, this plurality of conceptions also translates into an opportunistic use of different conceptions by the students to negotiate their role as students in classroom contexts. For instance, Nielsen (2012) found that in argumentations on socioscientific issues, students often shifted their interpretation of nature when their arguments were challenged, and their invocations of nature "were often uncritical appeals and rarely involved science factual content" (p. 723). As Cobern (2000) argues, these diverse conceptions are likely reflective of different perspectives such as religious, scientific, aesthetic, and conservationist and their experiences with the nonhuman world that children acquire over the course of their lives as students and participants in the social life outside school.

In addition to multiplicity of perspectives on nature, students also manifest diverse attitudes and values about nature that can perhaps be best captured by the following three recurring themes. First, students express positive emotions and attitudes about nature. Studies show that most students express pro-ecological worldviews and value nature for its material, aesthetic, and recreational values (Bozzolasco, 2017; Dai, 2011; Rickinson, 2001). They report enjoyment of nature and see it as a place for leisure activities and solitude (Cheng & Monroe, 2012; Rickinson, 2001). Second, they also feel fear and pessimism regarding future environmental health of the planet. Studies indicate that most students experience negative emotions, such as fear, sadness, and anger, when asked to share their feelings about environmental problems (Strife, 2012; Rickinson, 2001). Students in Ms. Gilmour's classroom, too, were pessimistic about future environmental conditions. When I asked a group of four students how the environmental problems that they had mentioned to me harm us in any way, the discussion proceeded as follows:

Zach: I think it is an outlook on the future ... the present ... when we think about problems ... it is about what is going to happen as it continues than what is happening in the next 5 or 10 years, but what is going to happen in 20 or 30 years in the future.
Sharma: OK, then who is going to be most affected by these problems?
Richard: Our grandchildren.
Zach: The next few generations.
Sharma: Okay ... so like you are talking about water pollution ... so you think the coming generation will get influenced?
Zach: Yeah.

Finally, a few studies also indicate that some students may perceive nature as a threatening space and associate it with danger and fear (Dai, 2011; Rickinson, 2001).

This diversity of views and attitudes about nature reveals a few interesting and important characteristics that distinguish how students come to perceive their world. Foremost among these would be the overwhelming tendency among students to exclude humans from their conception of the natural world. For example, Payne et al. (2014) in their study of sixth-grade Australian children found that "Most children conceived nature as living and non-living things existing naturally in the external environment.... Minimal human influence, interference or effect was identified as a primary characteristic of natural nature" (p. 70). Similarly, in the United States, Shepardson et al. (2007) found that students most commonly perceived the environment as a natural place where animals and plants live with the exclusion of humans. Such a perception is typically supported and affirmed by school science, because as we have discussed in the previous chapters, this nature-social dualism is characteristic of school science curricula, textbooks, and classroom discourse. In their conversations with me, Ms. Gilmour's students remarked that when they learned about ecology concepts like biomes, ecosystems, and food chains, humans were not shown as an integral or important component of these systems. For example, when I asked a group of students to list all the creatures that are present in a biome, the following exchanged ensued:

Sharma: In a biome what all creatures are present? Pick any biome.
Nick: In a savannah you will probably find wildebeest.
Sam: She (Ms. Gilmour) said that lots of other big animals are found there, like wildebeest, buffalos, and elephants and giraffes maybe. I don't know.
Sharma: What about human beings? What about us? Are we a part of any biome? Or are we not a part of any biome?
Mary: It kinds of depends upon where you live ... it kind of more like ... not really civilization there ... it was more like just some little villages there ... not a lot of big cities and stuff like that in a savannah.
Sharma: Okay? So did you talk about how human beings were part of Savannah?
Mary: Not really.
Sam: Not really.

As we see in this exchange, when asked to list creatures present in a biome, Ms. Gilmour's students tended to list animals and exclude humans. But, when specifically asked about the place of humans, they admitted that humans too were part of biomes. This indicates that rather than simply following a nature-social dualism, many students could be developing a more nuanced perception of humans in relation with nature. This perception allows them to consider humans as distinct from and yet a part of nature. Ms. Gilmour's students in this way are similar to a substantial number of 13–14-year-old Australian students in Pointon's (2014) study who too expressed a view of nature that was neither predominantly ecocentric (humans as integral part of nature and not different from other creatures) nor primarily anthropocentric (humans as completely separate and different from nature). Following Bonnett (2004), Pointon calls such a perception a "human-related" view of nature that "recognises the distinctive place humans inhabit in the world, both a part of and yet separate from nature whilst also recognising: the essential 'otherness' of nature; the integrity of nature; the continuity of nature and the intrinsic value of nature" (2014; p. 787). At least in this respect, these children are not different from many adults in these societies who likewise consider themselves as part of nature, but still when asked to describe nature may represent it as a space free of humans (Hoalst-Pullen, Lloyd, & Parkhurst, 2013; Vining, Merrick, & Price, 2008). On the whole, therefore, it seems fair to conclude that in the United States and similar advanced capitalist societies, though a majority of students do not consider humans as part of nature, there is a substantial number of students (and adults) with a nuanced and even contradictory view on whether humans are separate from or part of nature. However, underlying these differences is a clear agreement on human exceptionalism and the "otherness" of nature.

Such a belief in the unique and superior place for humans vis-à-vis other forms of life is well aligned with an accompanying belief that humans have a primarily utilitarian relationship with the nature. Thus, Pointon (2014) found a majority of the students in her study saw nature "primarily as a resource at the disposal of humankind" (p. 784). However at the same time, students also tend to see humans as stewards of nature and feel responsible for maintaining a livable environment for all species (Li & Ernst, 2015). As analyzed in the previous chapters, such views correspond well with the representation of human-nature relationships in school science curricula, textbooks, and classroom discourse. As is the case with other concepts in ecology, students' understanding of the connections between the human and natural world is usually week and patchy. For instance, Tsurusaki and Anderson (2010), in their study of students' understanding of connections between human engineered and natural environmental systems, found that students had a poor understanding of supply and waste disposal chains and environmental issues. So much so that many steps and processes in supply and waste disposal chains were not even visible to students. Alarmingly, this was particularly true for steps that have the greatest impact on our natural environment. For example, feedlots were invisible to the students in the beef supply chain, and they also had a poor understanding of how landfills work and fit in as part of the waste disposal chain. In my interviews with Ms. Gilmour's students, I too found that students had a poor understanding of the complexities of human dependence upon the rest of the planet. For instance, they primarily saw humans as participating in the natural world on an individual basis, such as individuals polluting the environment by throwing trash or exploiting and harming it by cutting down trees for their individual use. They never embedded human interaction with the biophysical world within larger socioeconomic and institutional or structural contexts and thus did not consider the role of social and economic systems in engendering and aggravating environmental problems.

Further, studies also report that students tend to have "an object view of nature, describing and depicting it as living things such as animals, trees, and plants that exist separate from other living factors and human beings" (Bozzolasco, 2017; p. 133). That is, when students are asked to describe nature in their own words, most students do not identify or highlight relationships or interactions between different inhabitants or components of the natural world and limit themselves to listing the different flora and fauna that in their view constitute nature (Bozzolasco, 2017; Loughland et al., 2002). Further, in close correspondence with school science representation of the natural world as a stable and well-ordered system, students likewise perceived nature as a "relatively static entity" (Rickinson, 2001; p. 276). For instance, Ergazaki and Ampatzidis (2012) found that most students in their study "found it very likely for a disturbed ecosystem to fully recover its initial state—mainly due to a 'recovery process' or inherent 'recovery mechanisms'—showing a strong belief in an extremely resilient 'Balance of Nature'" (p. 511). A comparable study with undergraduate students in the United States reported that most undergraduate students believe that a balance of nature exists for real ecological systems (Zimmerman & Cuddington, 2007). Again, like a belief in human exceptionalism, a belief in the balance of nature appears to be widely shared among students and adults alike (Cutler, Leiserowitz, & Rosenthal, 2017).

In the next section, we review students' understanding of environmental issues and what can be done to alleviate them. Readers will notice that students' understanding of these issues match well with their perceptions of and attitudes about nature. Further, just as we saw in the case of their perceptions of nature, we will see that their understanding of environmental issues correlates well with how adults view these matters. As we have asserted throughout this book, we see these correspondences primarily as a sign of the pervasive influence of a few discourses, such as the scientific and neoliberal discourses, that dominate both in science education and in the broader social life in advanced capitalist nations.

Students' Understanding of Environmental Issues

Ms. Gilmour spent a significant amount of time in her environmental science unit covering major environmental issues facing the planet, even though these topics weren't explicitly included in the state science standards for the grade (seventh) she was teaching. However, she mostly took a facilitator's role on this topic by letting students do research and present on an environmental issue of their choice. It was clear from their class presentations that by researching one topic of their choice, students had indeed learned a lot about their topic. But, on the whole, the impression I got by talking to them was that the factual environmental knowledge of most students was generally low. In this respect, these students may not be much different from students in other communities and even other nations (Rickinson, 2001). When I asked them about the environmental problems they were aware of, their responses were mostly short and limited to a few boilerplate issues covered in the mainstream media, such as pollution and deforestation, though students also mentioned greenhouse effect, oil spills, overpopulation, and desertification. When asked for elaboration, they were mostly comfortable talking about pollution and deforestation. These results also match what has been reported in other studies. For instance, air pollution emerged as the leading environmental problem in a study on seventh-grade students in Turkey (Özdem et al., 2014), while an older literature review on this issue also reported that water and air pollution were seen as the most serious environmental problems by students in the United States (Rickinson, 2001).

Interestingly, despite heavy coverage in the media for more than a decade, climate change does not yet seem to register as a prominent environmental concern among students at any grade level. Among Ms. Gilmour's students too, only one student indirectly cited climate change as a concern by mentioning the greenhouse effect. We did not explore the reasons for why this omission persists, and they are hard to infer from existing research. It is possible that students may be reluctant to mention climate change as an environmental concern because they generally lack a good understanding of how climate change operates and suffer from significant misconceptions regarding its causes and solutions (Liarakou, Athanasiadis, Gavrilakis, 2011; Shepardson, Niyogi, Choi, & Charusombat, 2009). It is also possible that they view climate change as a contested political issue that does not align with their broader worldview and that they may be critiqued for raising (Stevenson, Peterson, Bondell, Moore, & Carrier, 2014). Also missing from students' lists of environmental concerns are a few other environmental issues identified as most critical by the scientific community, such as loss of biodiversity and land degradation (Ehrlich & Ehrlich, 2013). These discrepancies may play a significant role in how students participate in democratic decision-making on environmental issues when they become adult members of their communities. Lacking good research, we cannot say much beyond stating that this is a topic that surely calls for greater sustained attention by the education research community.

It is also noteworthy that Ms. Gilmour's students did not mention environmental problems that were local in nature. It was only when I prodded them on this topic that they mentioned a few environmental issues that could be seen as local, though still rarely specific to their community. Consider these two exchanges I had with two groups of students:

(A) With Zach, Richard, Darrell, and Jonathan:

Sharma: Okay, then are you aware of any environmental problems that are local?
Zach: Well, we have been told in the past about how cars cause pollution ... and carbon and burning fuels and stuff ... But that's about it.

(B) With Beth, Melissa, Rex, and Gavin:

Sharma: Do you know about any local environmental problem?
Gavin: I guess pollution is an environmental problem everywhere. And also deforestation.
Rex: And there is something about this river. It is all trashed up.
Sharma: Which river are you talking about?
Students are unable to recall the name.
Beth: Around here I guess Lake Oconee is also polluted.

As can be seen in both cases, students' responses lack specificity and detail, and when local issues were raised, students tended to downplay their relative severity. For instance, when I asked Zach, Richard, Darrell, and Jonathan if they thought that deforestation was a local problem as well, Zach thus responded:

> Here too but maybe not as much as in other places ... At least a lot of people here have attempted to do stuff ... but there are many other places where there are not many other ways to make a living ... like in Africa ... there are not a ton of jobs and people have to make money ... so they cut trees when they don't have jobs.

These students' perceptions that environmental problems are bad things that primarily happen elsewhere were also reflected in the fact that in their class presentations on environmental problems, only three students chose a problem that was local in origin and nature. Two of them chose to focus on the ongoing dispute regarding sharing of the Chattahoochee River's waters, the so-called water wars between Georgia, Alabama, and Florida, while the third student gave a presentation on Kudzu—the vigorously invasive plant species common in much of the southeastern United States. The remaining students focused on distant, widespread, or global issues, such as rainforest destruction, coral reef decline, overfishing of the oceans, climate change and the polar bears, climate change and the penguins, invasive gypsy moth species, and water pollution. Research done in other societies also suggests that students see global and "foreign" environmental issues as more serious than local ones (Rickinson, 2001).

Further, students in Ms. Gilmour's classroom did not see themselves as being in any danger of direct harm from environmental crises. At the most, they saw future generations as possibly being at risk from these problems. For instance, Zach admitted that,

THE RECEIVED CURRICULUM: NATURE AS UNDERSTOOD BY STUDENTS 159

> It will become a problem if we don't get enough oxygen because trees make oxygen. ... the carbon dioxide that we breathe out, the trees and plants change that to ... that is how they produce oxygen. And when we burn all this gas and oil and stuff like that we put more pollutants in the air which can cause ... if not making breathing difficult, there would just less oxygen in the air. Again if not now ... the continuing of this for the next 20, 30, 40 years ... I mean it might be 100 years before we really, really see a huge difference at all ... I mean it is not affecting us now, but continuing it if nothing happens then it could be an issue.

When I quizzed Zach further about whether environmental problems that are not affecting people like him now might currently be affecting any other communities, he acknowledged the following possibility:

> In certain places like desertification ... in Africa a lot of nomadic people are overgrazing ... that has progressed actually very quickly like in 50 years there has been a huge amount of overgrazing. If it is overgrazed too much then it (the grass) goes away completely ... and that would certainly be a large problem for people down there.

That is, Ms. Gilmour's students mostly saw other animals that lived mostly in distant locations as those being at short-term risk of harm from worsening environmental conditions. For example, Rex opined

> People burn forests to make land for farms and cities ... and they don't think about what animals may go extinct ... they are just not thinking about it.

In their projects on environmental problems, these students portrayed the impacts of environmental problems largely in terms of habitat loss and threat of species extinction. It is quite probable that students' environmental concerns are more influenced by their socioeconomic profile and family contexts than by their school science experiences (Boyes, Skamp, & Stanisstreet, 2009; Grønhøj & Thøgersen, 2017).

Thus, the "othering" and displacement of environmental issues to remote locations serves a sociopolitical purpose of dampening students' environmental concerns and reducing their inclination to engage in environment-friendly practices that may have economic consequences (Barma, Lacasse, & Massé-Morneau, 2015). Again, this is an issue that calls for more urgent attention and research by the science education community. One potentially fruitful direction could be to study communities where environmental problems are nearly impossible to ignore, such as in those areas prone to or recovering from environmental disasters. For instance, Buxton (2006, 2010) found that students in hurricane-prone areas are enthusiastic about and committed to learning about the roles the humans play in exacerbating environmental problems. The string of natural disasters that afflicted the United States in 2017 provide a range of contexts for further study in this area.

We now come to the causes of environmental problems that were identified by Ms. Gilmour's students. The students overwhelmingly blamed individuals, collectively labeled as "people," for creating or worsening environmental problems. For example, see the following group dialogue:

Sharma: OK, what are the causes of pollution?
Rex: People.
Sharma: People?
Beth: Yeah, that is a big one. (We all laugh.)
Sharma: What kind of people? Like people like you and me or somebody else?
Gavin (speaking slowly emphasizing each syllable): B-A-D people.
Rex: Like everybody pollutes, though they might not notice it.
Sharma: Yeah.
Rex: They throw trash on the side of the road. ... That's polluting. They just don't realize it.
Sharma: Yeah, yeah ...
Rex: Like putting gum on the sidewalk and somebody steps on it.
Gavin: Or like people in the factories ... like smoke coming out of the smokestacks ... like cars even.
Rex: In Mexican cities sometimes you can't even go out because the smog is so bad.

Even in situations when students identified some other entity, such as a corporation, for harming the environment, there was a tendency to identify individuals as the primary agents for causing that harm. For instance, the conversation with Rex, Beth, Gavin, and Melissa later veered towards the Deepwater Horizon Oil Spill of 2010 and proceeded as follows:

Melissa: And that oil spill ...
Sharma: Yeah, that oil spill ... why did that happen?
Gavin: because the BP didn't do what they were supposed to.

Melissa: They should have really been careful.
Sharma: So is it BP's fault?
Gavin: It is not the company's fault ... it is the people that were working on that rig.

The tendency to blame individual people for damaging the environment also showed up in students' project reports on environmental problems. In almost all reports students explicitly or implicitly found ordinary individuals, often labeled as "people," culpable for causing environmental harm. For example, in a report on rainforest destruction, Cheryl and Heather listed logging as one of the main reasons for rainforest destruction and asserted that

> Many people cut down the trees in the rain forest for the wood, they are loggers that bought that land and they just destroy the land by cutting all the trees down. Also many people do illegal logging and people have been more and more successful catching them while they are doing it. Still today many people get away with it and destroy an important resource.

Such claims about the role of "people" clearly stand in stark contrast to the depressing accounts of destruction of rainforests, such as those in Borneo, by timber mafia and logging companies (Straumann, 2014). These nonlocal entities, often in complicity with global financial institutions, have not only laid waste to large swathes of healthy rainforests in Southeast Asia and many other parts of the world but have also caused incalculable harm to indigenous communities living in those regions (Dauvergne, 2016; Vandermeer & Perfecto, 2005). Other studies also indicate that students are rarely "critical of the effect of socio-economic structure on the environment" (Aguirre-Bielschowsky, Freeman, & Vass, 2012; p. 91).

We should not be surprised by such findings. In the previous chapter, we saw that the school science discourse in Ms. Gilmour's classroom did not differentiate between individual people and larger economic forces when it came to polluting or harming the natural world. Everyone came across as equally blameworthy. More importantly, the role that industrialized global systems of resource exploitation and consumption play in endangering the planet was rarely acknowledged. This occlusion, as we saw in the third and fourth chapters, is legitimized and encouraged by the mandated science curricula and textbooks. It is hardly surprising then that most students, as well as adults, remain ignorant about the role of industrialized supply and waste disposal chains in creating environmental crises (Tsurusaki & Anderson, 2010; Vandermeer & Perfecto, 2005).

When it comes to solving environment crises, studies show that students are most likely to focus on individual actions than on larger systemic, private sector or government-based solutions (Besel, Burke, & Christos, 2017). My conversations with Ms. Gilmour's students too showed a marked preference for virtuous individual-based environmental action. For example, when I asked what can be done about environmental problems, Melissa and Rex responded as follows:

Melissa: Donate money to help. Donate money maybe.
Sharma: Donate money, Okay.
Melissa: I guess you can also volunteer.
Rex: For the overall environment, I guess just recycling.

These students said they recycled at home, and several were members of Ms. Gilmour's Ecology Club, and thus, also helped in the school's recycling efforts. However, they could not adduce any other evidence or examples of environmental activism that they could perform. They were also skeptical of the impact that they could have at an individual level regarding some problems. For instance, talking about air pollution, Megan said "I mean there are things you can do like not throwing out your trash ... But as far as driving cars go and production in factories. These are some things that you kind of ... need." Here Megan appears to be echoing what Lorenzoni and Pidgeon (2006) found when adult participants in their study also acknowledged significant social barriers that inhibited their ability to act in environmentally responsive ways. Other studies have also shown that students tend to be less environmentally agentive on issues that are particularly associated with their own lives and material aspirations (Rickinson, 2001). Thus, it is hardly surprising to find that students' reports of undertaking environmentally responsive actions often fall short of their verbal commitments and feelings (McBeth & Volk, 2009).

There was also a belief among these students that the preferred way to solve these problems was through technological innovations that could replace the need for people to change their behaviors. Thus, when we were discussing if anything can be done about the problem of pollution, Richard, Darrell, and Jonathan made the following suggestions:

Richard: You can make cars that are ...
Darrell: Solar powered?
Jonathan: Yeah ... or cars that use gas made out of vegetables.

A trust among students in the ability of science and technology to solve environmental problems has been noted in other studies as well, such as Liu and Lin (2014). As we saw in our analysis of science standards, this view is encouraged by official science curricula. It is also increasingly reflected in public policy owing to the dominance of ecological modernist views among policymakers (Dryzek, 2013) and finds widespread acceptance among adults in developed nations of the West (Leiserowitz 2007; Lorenzoni, Nicholson-Cole, & Whitmarsh, 2007). Given the fact that students' views are shaped by a combination of out-of-school and in-school science discourses, we view the close correspondence among dominant ecological discourses as a major reason why students and adult citizens are roughly on the same page as the policy elites in the United States when it comes to environmental problems and solutions.

Summing Up

In this chapter, we integrated results from our analysis of conversations with Ms. Gilmour's students with existing research on environmental issues to reach some conclusions about how students perceive nature and environmental problems. However, because the research on this issue, especially with US students, remains patchy and sporadic, we could not achieve more than a tentative understanding of students' views on this issue. We have presented it as such, with an eager anticipation that our work will contribute to greater attention and more sustained research on this topic in the United States and perhaps in other parts of the world as well.

Our analysis indicates that despite possessing limited scientific knowledge and understanding of ecology, students come to acquire distinct perceptions, attitudes, and values about our biophysical world and our role in it as humans. Further, though they display diverse conceptions and attitudes about the natural world, students' views are marked by a belief in human exceptionalism and the "otherness" of the rest of the environment. This "othered" world is a stable human-free world that is understood more in terms of entities inhabiting it than in terms of the relationships and underlying processes that sustain it. Further, nature seen in this way may be appreciated by these students for its aesthetic and recreational values, but our relationship with nature is largely constructed in instrumental terms.

Students realize that all is not well with this world, but our results show that the students in the United States feel that at least for now and in near future, they and people like them are largely safe from the harmful consequences of environmental damage to the planet. They may also realize that other animals and perhaps people too in some distant communities are facing grave threats from environmental crises, but these crises seem to them to be both foreign, and perhaps, inevitable. Students seem to blame ordinary people for these problems and do not yet have an appreciation of the larger systemic causes that limit the scope of individual culpability and environmental action. As in other nations, students in the United States are largely pro-environment in their outlook but are unable to think of ways to heal the planet that go beyond individual "green" actions, such as recycling. Consequently, they look to advances in science and technology to solve these problems. On the whole, it appears that students are not that different from adults when it comes to their views on nature and their relationships with the environment. Further, their views seem to mirror how school science as well as other dominant societal discourses construct the natural world and our relationship with it. This congruence leads us to conclude that, at least in the current circumstances, it may not be easy for alternative views on environmental issues to create lasting change in students' understandings on the matter.

Note

1. Hereafter referred to in the first person.

References

Aguirre-Bielschowsky, I., Freeman, C., & Vass, E. (2012). Influences on children's environmental cognition: A comparative analysis of New Zealand and Mexico. *Environmental Education Research, 18*(1), 91–115.

Allen, M. (2014). *Misconceptions in primary science*. Berkshire, UK: McGraw-Hill Education.

Barma, S., Lacasse, M., & Massé-Morneau, J. (2015). Engaging discussion about climate change in a Quebec secondary school: A challenge for science teachers. *Learning, Culture and Social Interaction, 4*, 28–36.

Besel, R. D., Burke, K., & Christos, V. (2017). A life history approach to perceptions of global climate change risk: Young adults' experiences about impacts, causes, and solutions. *Journal of Risk Research, 20*(1), 61–75. https://doi.org/10.1080/13669877.2015.1017830

Bonnett, M. (2004). Lost in space? Education and the concept of nature. *Studies in Philosophy & Education, 23*(2/3), 117–130.

Boyes, E., Skamp, K., & Stanisstreet, M. (2009). Australian secondary students' views about global warming: Beliefs about actions, and willingness to act. *Research in Science Education, 39*(5), 661–680.

Bozzolasco, A. M. (2017). *Using a mixed-methods approach to understand urban children's nature conceptions, ecological worldviews, and environmental perceptions and preferences before and after attending an environmental education program*. Montclair State University.

Buxton, C. (2006). Creating contextually authentic science education in a "low performing" urban elementary school context. *Journal of Research in Science Teaching, 43*(7), 695–721. https://doi.org/10.1002/tea.20105

Buxton, C. (2010). Social problem solving through science: An approach to critical place-based science teaching and learning. *Equity and Excellence in Education, 43*(1), 120–135. https://doi.org/10.1080/10665680903408932

Cheng, J. C. H., & Monroe, M. C. (2012). Connection to nature: *Children's affective attitude toward nature. Environment and Behavior, 44*(1), 31–49.

Cobern, W. W. (2000). *Everyday thoughts about nature: A worldview investigation of important concepts students use to make sense of nature with specific attention to science*. Dordrecht, The Netherlands: Springer.

Covitt, B. A., Tan, E., Tsurusaki, B. K., & Anderson, C. W. (2009). *Students' use of scientific knowledge and practices when making decisions in citizens' roles*. Paper presented at the annual conference of the National Association for Research in Science Teaching Garden Grove. http://edr1.educ.msu.edu/EnvironmentalLit/publicsite/html/report_2009.html

Cutler, M., Leiserowitz, A., & Rosenthal, S. (2017). *Is nature stable, delicate, or random?* Retrieved from http://climatecommunication.yale.edu/publications/nature-stable-delicate-random/

Dai, A. H.-I. (2011). *A study of Taiwanese children's conceptions of and relation to nature: Curricular and policy implications*. College Park, MD: University of Maryland.

Dauvergne, P. (2016). *Environmentalism of the rich*. Cambridge, MA: MIT Press.

Dryzek, J. S. (2013). *The politics of the earth: Environmental discourses*. Oxford, UK: Oxford University Press.

Ehrlich, P. R., & Ehrlich, A. H. (2013). Can a collapse of global civilization be avoided? *Proceedings of the Royal Society B: Biological Sciences, 280*(1754).

Eilam, B. (2012). System thinking and feeding relations: Learning with a live ecosystem model. *Instructional Science, 40*(2), 213–239.

Elliot, G. (2016). *Exploring nature as representation and young adults' conceptualisations of nature in the user-generated online world: Nature 2.0*. University of Otago.

Ergazaki, M., & Ampatzidis, G. (2012). Students' reasoning about the future of disturbed or protected ecosystems and the idea of the "balance of nature". *Research in Science Education, 42*(3), 511–530.

Grønhøj, A., & Thøgersen, J. (2017). Why young people do things for the environment: The role of parenting for adolescents' motivation to engage in pro-environmental behaviour. *Journal of Environmental Psychology, 54*, 11–19.

Hoalst-Pullen, N., Lloyd, M. R., & Parkhurst, M. E. (2013). Environmental attitudes and perceptions: A comparison of Peru and the United States. *Journal of Global Initiatives: Policy, Pedagogy, Perspective, 7*(2), 12.

Jordan, R., Gray, S., Demeter, M., Lui, L., & Hmelo-Silver, C. E. (2009). An assessment of students' understanding of ecosystem concepts: Conflating ecological systems and cycles. *Applied Environmental Education and Communication, 8*(1), 40–48.

Kellert, S. R. (2002). Experiencing nature: Affective, cognitive, and evaluative development in children. In P. H. Kahn & S. R. Kellert (Eds.), *Children and nature: Psychological, sociocultural, and evolutionary investigations* (pp. 117–151). Cambridge, MA: MIT Press.

Leiserowitz, A. (2007). *Human development report 2007/2008: International public opinion, perception, and understanding of global climate change*. Retrieved from http://hdr.undp.org/en/reports/global/hdr2007-2008/papers/leiserowitz_anthony6.pdf

Li, J., & Ernst, J. (2015). Exploring value orientations toward the human–nature relationship: A comparison of urban youth in Minnesota, USA and Guangdong, China. *Environmental Education Research, 21*(4), 556–585. https://doi.org/10.1080/13504622.2014.910499

Liarakou, G., Athanasiadis, I., & Gavrilakis, C. (2011). What Greek secondary school students believe about climate change? *International Journal of Environmental and Science Education, 6*(1), 79–98.

Liu, S., & Lin, H. (2014). Undergraduate students' science-related ideas as embedded in their environmental worldviews. *International Journal of Science and Mathematics Education, 12*(5), 1001–1021.

Lorenzoni, I., Nicholson-Cole, S., & Whitmarsh, L. (2007). Barriers perceived to engaging with climate change among the UK public and their policy implications. *Global Environmental Change Part A: Human & Policy Dimensions, 17*(3/4), 445–459.

Lorenzoni, I., & Pidgeon, N. (2006). Public views on climate change: European and USA perspectives. *Climatic Change, 77*(1), 73–95.

Loughland, T., Reid, A., & Petocz, P. (2002). Young people's conceptions of environment: A phenomenographic analysis. *Environmental Education Research, 8*(2), 187–197.

McBeth, W., & Volk, T. L. (2009). The national environmental literacy project: A baseline study of middle grade students in the United States. *The Journal of Environmental Education, 41*(1), 55–67.

Nielsen, J. A. (2012). Arguing from nature: The role of 'nature' in students' argumentations on a socio-scientific issue. *International Journal of Science Education, 34*(5), 723–744.

Özdem, Y., Dal, B., Öztürk, N., Sönmez, D., & Alper, U. (2014). What is that thing called climate change? An investigation into the understanding of climate change by seventh-grade students. *International Research in Geographical and Environmental Education, 23*(4), 294–313.

Payne, P., Cutter-Mackenzie, A., Gough, A., Gough, N., & Whitehouse, H. (2014). Children's conceptions of nature. *Australian Journal of Environmental Education, 30*(1), 68.

Pointon, P. (2014). 'The city snuffs out nature': Young people's conceptions of and relationship with nature. *Environmental Education Research, 20*(6), 776–794. https://doi.org/10.1080/13504622.2013.833595

Rickinson, M. (2001). Learners and learning in environmental education: A critical review of the evidence. *Environmental Education Research, 7*(3), 207–320.

Shepardson, D. P., Niyogi, D., Choi, S., & Charusombat, U. (2009). Seventh grade students' conceptions of global warming and climate change. *Environmental Education Research, 15*(5), 549–570.

Shepardson, D. P., Wee, B., Priddy, M., & Harbor, J. (2007). Students' mental models of the environment. *Journal of Research in Science Teaching, 44*(2), 327–348.

Stevenson, K. T., Peterson, M. N., Bondell, H. D., Moore, S. E., & Carrier, S. J. (2014). Overcoming skepticism with education: Interacting influences of worldview and climate change knowledge on perceived climate change risk among adolescents. *Climatic Change, 126*(3–4), 293–304.

Straumann, L. (2014). *Money logging: On the trail of the Asian timber mafia*. Basel: Bergli Books. Schwabe AG.

Strife, S. J. (2012). Children's environmental concerns: Expressing ecophobia. *The Journal of Environmental Education, 43*(1), 37–54.

Tsurusaki, B. K., & Anderson, C. W. (2010). Students' understanding of connections between human engineered and natural environmental systems. *International Journal of Environmental and Science Education, 5*(4), 407–433.

Vandermeer, J. H., & Perfecto, I. (2005). *Breakfast of biodiversity: The political ecology of rain forest destruction*. Oakland, CA/New York, NY: Food First Books.

Vining, J., Merrick, M. S., & Price, E. A. (2008). The distinction between humans and nature: Human perceptions of connectedness to nature and elements of the natural and unnatural. *Human Ecology Review, 15*(1), 1.

Zimmerman, C., & Cuddington, K. (2007). Ambiguous, circular and polysemous: Students' definitions of the "balance of nature" metaphor. *Public Understanding of Science, 16*(4), 393–406. https://doi.org/10.1177/0963662505063022

CHAPTER 7

A Sustainability Science-Based Framework for Science Education

By all accounts a majority of US adults are concerned about the environment and rank environmental protection as a top priority (Anderson, 2017). However, as things stand, very few adults in the United States actually do anything to act on those concerns beyond recycling, or even possess the basic knowledge necessary for well-informed, evidence-based environmental action (Anderson, 2017; Assaraf & Damri, 2009; Lorenzoni & Pidgeon, 2006). At the societal level too, the United States can't seem to decide upon a steady and sustained course of action that would meet the grave environmental challenges that face our planet. Thus, while some states, such as California, are forging ahead with ever more rigorous environmental policies, we see a dangerous decline in the federal government's commitment to tackle environmental crises as abundantly reflected in the Trump administration's decision to withdraw from the Paris Climate Accord and roll back the previous administration's Clean Power Plan (Davenport & Nagourney, 2017; Puko, 2017; Shear, 2017). Clearly, the current situation is untenable, and it does not appear hyperbolic to suggest that the need for forward-looking environmental action has never been greater. As science educators, we naturally see science education both as a field that must take a leadership role in promoting environmental action and where we can make the most meaningful contribution.

Thus, though educationalizing of societal problems has been subjected to criticism by some intellectuals, such as Fendler (2008) and Labaree (2008), we remain of the view that efforts to improve science education in

© The Author(s) 2018
A. Sharma, C. Buxton, *The Natural World and Science Education in the United States*, https://doi.org/10.1007/978-3-319-76186-2_7

schools do indeed play an important role in preparing our society to survive and perhaps even thrive in the current age of climate change and a sixth mass extinction. There are, of course, many ways to improve science education to meet this challenge. One of which is to examine and suggest changes in the "science" that is being taught in the schools, and that became the theme of this book. By choosing this path, we have attempted to join a select group of researchers that include Tasos Hovardas, Konstantinos Korfiatis, Elizabeth Hufnagel, John Ruppert, and Noah Feinstein. Our goal has been to further strengthen their case for bringing science education in sync with the current state-of-the-art understanding among ecologists and social scientists. Given our methodological home base of critical discourse analysis, we choose to make our contribution by ways of (a) a critical, systematic, and multidimensional examination of science education as a key site in the social and discursive production of nature and our relationship with it, and (b) positioning sustainability science as the overall framework for bringing school science in consonance with ecological and social sciences.

In this concluding chapter, we begin with a review of the main themes that emerged from our analyses of how natural systems and their relationships with humans are represented in science standards, curriculum material, and school science discourse. We organize these themes in terms of (a) what was mandated through standards documents and textbooks, (b) what was taught, and (c) finally what was learned and understood by the students. We next summarize the nature of understanding that students develop as a result of science instruction as well as out-of-school influences. Then, we present a few key attempts that have been made to improve how we teach about nature in K-12 education. Our review of such efforts is not intended to be comprehensive but illustrative so as to give the readers an idea of the range of ongoing efforts in this direction. The review is followed by our proposal for change that we think is needed to bring science education in line with the latest developments in the fields of ecology and social sciences. We conclude the chapter by drawing key implications of our proposal for curriculum developers, teachers, students, broader community, and future research efforts on this issue.

NATURE IN SCHOOL SCIENCE: THE INTENDED CURRICULUM

As in other subjects, science standards and textbooks as the intended curricula represent the "official" content knowledge in K-12 settings. Using critical discourse analysis, in Chap. 3 we examined assumptions, values, and perspectives about the natural world embedded in the relevant ecology, environmental, and earth science standards in the *Next Generation Science Standards* (NGSS) and two sets of science standards from the state of Georgia: the recently phased out *Georgia Performance Standards* (GPS) and the new incoming *Georgia Standards for Excellence* (GSE). In addition, in Chap. 4, we also examined the representations of the natural world and its relationship with humans in ecology and environmental science-related sections of a middle school science textbook *Georgia: Holt Science and Technology: Life Science* that is followed in many public schools in the state of Georgia. Our analysis shows that the science standards embedded within these curricular documents embody certain key assumptions that are further supported by a distinct set of values about the world. These assumptions and values combine to constitute some important perspectives on our world. These perspectives are enacted through a few well-aligned discourses that are currently dominant in science education as well as in social life. The science standards and textbooks work to naturalize these perspectives through their representation of nature and its relationships with the social world. We summarize our findings as follows.

Ontology: What Is There?

Children develop their understanding of the world around them by labeling things they see, sorting and differentiating entities into categories. They associate objects and experiences with meanings or interpretations through chains of signification and by ascribing relationships between the labeled entities. This emerging ontology is a critical component of their development because as Gardner (2011) asserted, "… the way in which children come to think of classes of entities affect the kinds of theories they develop about these classes and the kinds of inferences they are prepared to draw" (p. 94). For example, if a student categorizes all flying creatures as birds, she will naturally tend to see a bat as a bird, ascribe all bird-like behavior to it, and will then relate to it accordingly. Students' lives and experiences outside school give them enough wherewithal to develop remarkably robust naïve ontologies of nature. However, it cannot be denied that science education also plays a role in shaping their ontology of nature, particularly in its capacity as a socially sanctioned way through which the society seeks to influence how children come to understand the world and figure their place in it (Driver, Asoko, Leach, Mortimer, & Scott, 1994).

Our analysis showed us that the science standards present a world to the students that is ontologically divided into two distinct yet related domains—a natural world that is devoid of humans and a social world comprising humans. Our analysis, thus, aligns with Hufnagel, Kelly, and Henderson's (2017) critique of the NGSS standards in which they also found that in these standards "environment is constructed as an entity separate from people through both exclusion and ambiguity" (p. 1). In the Georgia GPS and GSE science standards as well as in the NGSS, it is assumed that the natural world can be understood as a system with interrelated and interdependent parts which interact through a combination of biological, chemical, and physical processes to exhibit emergent properties. Though nature as a system can change, the norm for it is represented as stable. If change occurs it is to be understood on the basis of a linear singular model of ecological succession. As for the social world, the standards make the assumption that humans have significantly impacted the natural world, but these texts do not discriminate between different communities, groups, or societies. That is, all human beings are lumped together under one category as "humans" with the implicit assumption being that all individuals can be seen as equally responsible for the "human impact" on the natural world.

Our analysis of the middle school science textbook *Georgia: Holt Science and Technology: Life Science* indicates that through a strategic use of grammar, these texts function as a technology being used to construct certain distinct representations of nature and its relationship with the social world for the students as well as the teachers. By and large, these representations support the "official" ontology of the world as inscribed in the science standards. Thus, for instance, we found that the *Georgia: Holt Science and Technology: Life Science* textbook supports the nature-social dualism of the science standards by presenting an ecological account of the natural world in which a vast majority of the related clauses made no reference to human beings or the social world. The only exceptions are the mental process clauses about humans thinking or feeling about some issues related to the natural world, and the material clauses that represented humans as scientists studying the natural systems.

Further, the textbook uses grammar to obfuscate human agency in nature-social relations either by suppressing human agency in those relationships or by attributing that agency to some anonymous, amorphous, and nonindividuated group, labeled simply as "people" or "humans." Thus, the overall impression given to teachers and students is that while it is important for them to understand environmental phenomena, it is not particularly relevant to know the specifics of human involvement in creating, modifying, or sustaining those phenomena. In addition, by attributing the agency for creating or aggravating environmental problems to "people," the textbook makes it appear as if ordinary laypersons are largely culpable for these problems. Finally, our textbook analysis shows that the official ontology of the world, as inscribed in the written curricula, externalizes the environmental threat by presenting animals and plants, but not humans, as the most likely victims of environmental problems. This can be seen as a way to sanitize environmental problems by omitting any mention of the devastating impact that environmental degradation is already having on marginalized communities around the world.

Values: What Should Be?

We examined the values embodied in the science standards by specifically focusing on those standards that related to human action and human reason. With regard to human action, it is evident that each of the three standards documents we analyzed take a negative view of the human impact upon the natural world. At the same time, however, there is a firm belief in the ability of humans to come up with solutions to environmental problems and an expectation that students will learn to take environmental action to reduce their impact on the natural world. Successful environmental actions inevitably have both technological and social components. However, in these standards documents we see a complete marginalization and hence devaluation of the social aspects of environmental action—an outcome that has also been reported by Hufnagel et al.'s (2017) critical discourse analysis of NGSS standards.

The science standards also project a distinct set of values around human reason. They present school science as not just a preparation for understanding the world as scientists do, but also as preparation to solve problems like them. By highlighting engineering ideas and practices in both NGSS and GSE, we see a strong preference emerging for a technocentric value system in these latest sets of standards, though technocentric values aren't absent from the earlier GPS either. Thus, science standards dissociate the social from the natural by scrubbing off all social aspects from environmental topics and issues. As a result, environmental issues are reduced to technological problems. Further, there is a clear positive valuation of the role of science and engineering in solving such issues. Our results echo Hufnagel et al.'s (2017) analysis of NGSS that showed that in this standards document, "when solutions to environmental issues are included the focus is on technoscience, sidelining the important and relevant social and political aspects of these problems" (p. 1). We also notice a tendency in these standards to fall back on an instrumental rationality for solving such problems, which gets manifested in a focus on understanding *process* over *cause* and *action* over *actors*, leading to valuing "how" questions over "why" questions for solving problems.

Along with technocentrism, we find that these documents also privilege an economic rationality in standards that deal with our relationship with rest of the world. That is, the intended curricula represent the human-nature relationship as one in which nature is valued either as a resource that we ought to be sustainably exploiting for meeting our needs or as a recipient of harm that we cause to it on account of activities that are primarily economic in nature. The marginalization of the social and the collective dimension of human action and privileging of technocentric and economic values shows that there is nothing objective about the representation of the world encoded in the official school science. In fact, a case can be made that these tacit and unexamined values, assumptions, and implications in the written curricula act in powerful ways to naturalize certain ideas and representations about our world in ways that support a few sociopolitically dominant perspectives on the world.

Perspectives on the World: Which Ideas Dominate?

Discourses as distinct forms of language in use offer definite perspectives on the world. Thus, through an iterative process of repeated reading and textual analysis of science standards, we identified "the main parts of the world (including areas of social life) which are represented – the main themes" and "the particular perspective or angle or point of view from which they are represented" (Fairclough, 2003; p. 129). We found four major themes in the science standards that pertain to (a) the ontological separation of the human and "natural" world; (b) an ecosystem ecology-based conceptual framework; (c) the focus of the scientific gaze upon the world; and (d) technocentric-economic rationality in human-nature relationships. Further analysis coalesced these emergent themes into two interrelated discourses, a *scientific discourse* and an *environmental discourse*, that appeared to have the most influence in shaping these standards documents insofar as representations of the world and our relationships with it are concerned.

Scientific Discourse Written by scientists and science educators, the science standards and textbooks present a Cartesian-Newtonian mechanistic perspective on the world that reifies the nature-social dualism and tells teachers and students that the world is out "there" as a physical reality that can be objectively observed, measured, modeled, and understood using scientific ideas and tools. Further, the scientific discourse in these documents represents the world as a hierarchically organized and well-ordered system of interacting biotic and abiotic components that exchange matter and energy while performing different underlying processes that yield observable characteristics and phenomena. Thus, ecosystems are presented as the appropriate units for investigating and understanding fundamental ecological processes and phenomena in the natural world. This is especially noticeable in standards that relate to matter cycles and energy flows. Following the cybernetic systems perspective of ecosystem ecology, curriculum writers expect students to understand the relationships between organisms primarily in terms of cycling of matter and energy flows that occur between them. The mechanistic Cartesian-Newtonian perspective of the scientific discourse in official school science meshes well and supports the other discourse—the environmental discourse—that we saw running through these documents.

Environmental Discourse We found that the environmental discourse in school science curricula is a hybrid discourse that borrows critical elements from two different existing discourses—*Ecological Modernization* and *Green Governmentality*. As a legitimizing discourse for capitalism in the age of environmental anxieties, *Ecological Modernization* promises a "green" capitalism that offers both continued economic development and environmental sustainability. Guided by an overall economic rationality, this discourse conceptualizes nature as a repository for meeting human needs and tells students that the natural world can and should be sustainably managed with the help of scientific knowledge and technological tools for meeting human needs. The discourse of *Ecological Modernization* is supported by the other discourse of *Green Governmentality* in these documents. *Green Governmentality* concurs with *Ecological Modernization* in its technocratic, managerial approach to nature and environmental threats. But its central focus is on the governance of individual and social life in matters related to our relationship with the natural world. It works by presenting local and individual environmental action as the only available option for individual citizens interested in preserving the environment. *Green Governmentality* backgrounds society's economic, political, and social systems as potential factors impacting nature-human relationships and works to orient students' attention towards population growth and individual consumption as the main reasons for the damage to the natural world. Thus, this discourse naturalizes a perspective in which the undemocratic governance structures and "green" capitalism are absolved of responsibility for ushering in the Anthropocene Epoch; while students are led to think that since it is they as individuals who are responsible for causing environmental damage, their environmental actions and choices (along with technological solutions) are needed to undo the harm.

Our findings, therefore, support Feinstein and Krischgasler's (2015) analysis of how sustainability is embodied in the Next Generation Science Standards. They found that the discourse around sustainability in these documents is marked by three dominant themes of *universalism, scientism,* and *technocentrism* that together evoke the discourse of *Ecological Modernization*. As a result, sustainability in NGSS is presented as "a set of global problems affecting all humans equally and solvable through the application of science and technology" (p. 121). We agree with them when they argue that "students who are taught to think about sustainability from this perspective will be less able to see its ethical and political dimensions and less prepared for the political realities of a pluralist, democratic society that must balance the needs of multiple groups and integrate science with other sources of knowledge to develop contextualized responses to sustainability challenges" (p. 121).

Cartesian-Newtonian mechanistic *Scientific Discourse* and the environmental discourses of *Ecological Modernization* and *Green Governmentality* are indeed the dominant official discourses that shape the intended curricula on topics related to the natural world and our relationship with it. As dominant discourses, they establish a regime of truth that works to naturalize and universalize a distinct set of assumptions, values, and representations about our world. Science educators need to critique this regime of truth to understand if the world so labeled and produced by these discourses is indeed the world that we and future generations would like to live in. For instance, we need to examine what kind of ontology of the world we are leading students to accept when we orient them to relate to the world through a rationality that is overwhelmingly instrumental and economic. Are we naturalizing for them the profound transmutation of entities and relations into commodities that can be extracted, transported, and traded in a global market? Of course, progressive commodification of our world has been happening since the beginning of the industrial revolution. Capitalist modes of production and consumption were always undergirded by the resignification and partitioning of our world into "fictitious commodities" (Polanyi, 1957). What is definitely new in the current neoliberal era is the naturalization of a conservation-as-development paradigm of green or natural capitalism (Büscher, Dressler, & Fletcher, 2014). This paradigm seeks to lull us into a dangerous complacency with the false promise of sustainable yet unlimited development through terraforming of the Earth. Our analysis of the intended curricula indicates that the science standards and textbooks are supportive of this paradigm and marginalize alternative ways of understanding and relating with the world. To understand if this effort is indeed successful we also examined the enacted curriculum of what was taught and the received curriculum, or what students actually understand about the natural world. We summarize what we found in the next two sections.

Nature in School Science: The Enacted Curriculum

In order to understand the nature of the taught, or enacted, curriculum on topics connected with the natural world, we did an ethnographic case study of how one teacher represented nature while teaching seventh-grade science in her classroom (see Chap. 5). Like most science teachers, this teacher, Ms. Gilmour, was seeped in the traditional scientific discourse that positioned scientists and official sources of knowledge, such as science textbooks, as presenting authoritative accounts of what the world is like. Ms. Gilmour believed in the mainstream authoritative scientific account of the natural world, and as we show in the case study, she worked hard to authoritatively disseminate it among her students. Thus, following the state science standards, she sought to present a Cartesian-Newtonian mechanistic perspective on the world to her students. The world was represented through a dualism that separated the natural from the social world with a clear understanding that the purview of science is restricted to the discursively created "natural world" alone. Further, the school science discourse in her classroom represented the natural world as an abstract system of conceptually dense terms that were linked together in a hierarchically organized network of relationships. Though Ms. Gilmour tried to bring in illustrative examples of abstract ecological ideas and processes, the abstract ontology of nature appeared removed from students' lived experiences with the world.

The representation of nature as an abstract system was accompanied by a compatible view of the natural world as a largely stable world that supported the "balance of nature" view that students routinely encounter in media, in everyday discourses, and that they are likely to see in future science classes. This abstract, stable world was made more amenable to understanding through ecosystem-based cybernetic models of energy and matter fluxes and by externalizing humans from the natural world. That is, humans were largely positioned as the external harmful influence on the natural world. We are afraid that this representation of our world lays the foundations for an incorrect and ecologically harmful lifelong orientation towards the place of humans in nature.

Scientific and environmental discourses come across as highly authoritative to participants who are marginal to its production. Science teachers belong to this category of peripheral members in the wider scientific discursive community even though they play a vital role in disseminating this discourse in society. Scientific discourse, especially in school settings, also tends to be a hegemonic discourse as it actively delegitimizes discourses deemed unscientific and normalizes certain perspectives and representations of our world for both teachers and students. Further, decades of reduced teacher autonomy and heightened teacher accountability in the United States have much reduced the incentives for teachers to teach anything other than the officially authorized ontology of the world. Thus, it is hardly surprising that Ms. Gilmour taught in ways that hegemonize the authoritative discourses of the science standards in the discursive space of her classroom. We have every reason to believe that Ms. Gilmour's teaching is a good representation of current science teaching at the K-12 level in the United States. Of course, because of the paucity of research on this issue, especially in the United States, this assertion needs to be further validated with more research on how science teachers represent the natural world to their students.

Nature in School Science: The Received Curriculum

Fortunately, there is substantially more research, albeit much of it is international in nature, on students' understanding of nature. Thus, we were able to braid existing research findings with patterns that emerged from our conversations with students in Ms. Gilmour's classroom to arrive at a more comprehensive and detailed view of the received curriculum as regards nature in school science (see Chap. 6). On the whole, we see a huge transmission loss between the intended curriculum, the enacted curriculum, and the curriculum that is received, or learned by the students, though, as we elaborate, this transmission loss is largely limited to the understanding of science concepts. The overall perspectives on the world as embedded in the scientific and environmental discourses of the science standards, on the other hand, manage to seep into the received curriculum with remarkable fidelity. Thus, as other research points out, and as we also discovered, students on the whole have a poor scientific understanding of the biophysical world which they label interchangeably as "nature" or "environment." What is even more sobering is that these students tend not to use whatever little scientific knowledge they do possess when making decisions in their roles as citizens.

However, it is also clear that students subscribe to diverse conceptions of nature, ranging from nature as a space for animals and plants to live, to nature as an inclusive domain where one finds all forms of life, including humans, in relationships of mutual sustainability and interdependence. In addition to this multiplicity of perspectives on nature, students also manifest diverse values about nature that are largely marked by positive emotions and attitudes mixed with fear and pessimism regarding the future environmental health of the planet. Some students also perceive nature as a threatening space and associate it with danger and fear. Despite the diversity of views and attitudes about nature, most students tend to exclude humans from their conception of the natural world. As is clear from the preceding discussion, such a perception is affirmed and supported by both the intended and the enacted curriculum. Further, in close correspondence with school science representation of the natural world as a stable and well-ordered system, students likewise perceived nature as a relatively static entity.

For some students, the relationship between nature and humans goes beyond a simplistic nature-social dualism, and these students come to acquire a more nuanced perception of humans in relation with nature. This perception allows them to consider humans as distinct from and yet a part of nature. However, despite different takes on nature-social relationship, most students believe in human exceptionalism and the "otherness" of nature. Such a belief in the unique and superior place for humans vis-à-vis other forms of life is well aligned with an accompanying belief that humans have a primarily utilitarian relationship with nature. Though, at the same time, students also tend to see humans as stewards of nature and feel responsible for maintaining a livable environment for all species.

Despite these pro-environment views, students' awareness of environmental problems is mostly superficial and limited to a few boilerplate issues covered in the mainstream media, such as pollution and deforestation. Interestingly, despite heavy coverage in the media for more than a decade, climate change does not yet seem to register as a prominent environmental concern among students at any grade level. Given the severe threat posed by climate change to human societies, the marginalization of climate change as an expressed environmental concern among students is indeed worrisome and certainly worthy of further research and consideration by science educators. In students' views we also see a tendency to externalize environmental threat by seeing these threats as bad things that happen to other species or other people living in poor and distant societies. Further research is needed to see if this "othering" and displacement of environmental issues to remote locations is serving a sociopolitical purpose of dampening students' environmental concerns and reducing their inclination to engage in environment-friendly practices that may have economic consequences.

In addition to externalization of environmental threat, we also see an individualization of environmental culpability in students' views. That is, ignoring larger systemic, societal causes, students tend to blame individuals, collectively labeled as "people," for creating or worsening environmental problems. Further, when it comes to solving environment crises, students are more likely to focus on individual actions than on larger systemic, private sector or government-based solutions. As in other nations, students in the United States are largely pro-environment in their outlook but are unable to think of ways to heal the planet that go beyond individual "green" actions, such as recycling. Consequently, they look to advances in science and technology to solve these problems. On the whole, it appears that students are not that different from adults when it comes to their views on nature and their relationships with the environment. Further, in these views we see an unmistakable influence of the environmental discourses of Ecological Modernization and Green Governmentality embedded in intended and enacted curricula, along with a strong compatibility with other influential societal discourses, such as conservative Christianity and neoliberalism (Büscher et al., 2014; Sherkat & Ellison, 2007). This congruence leads us to suspect that sporadic or piecemeal efforts may not be enough to prepare students for building a sustainable, equitable world in the Anthropocene Epoch. This is especially so in the current sociopolitical climate, in which powerful groups and individuals interested in preserving status quo have deployed substantial material and discursive resources to delegitimize or marginalize science as a source for valid and reliable knowledge on environmental issues (Dryzek, 2013; Oreskes & Conway, 2011). But at the same time, we also believe that it is extremely important for science educators to continue with the limited reform-oriented work they can do in their own contexts and networks of relations so that when an opportune time comes, diverse piecemeal efforts can come together to contribute to a successful reimagining of science education for the current era. In the next section, we outline two such efforts that have attracted attention or shown promise in the context of education in the United States.

Promising Practices for Reforming Environmental Education

Typical educational reform efforts germinate at the top levels of education policymaking and then gradually take concrete shape in terms of programs and mandates that filter down to classroom settings. But ideas for educational change can and do also germinate at school and classroom levels and spread beyond their initial settings to cast a wider influence on education. In this section, we will talk about two proposals for reform that began at very different places but are currently spreading in all directions to shape science education in the United States.

Education for Sustainable Development

The Earth Summit, held in Rio de Janeiro in 1992, was a watershed moment in global efforts to move the world towards a more sustainable path to development. It led to an important climate change convention that some years down the line was instrumental in the successful conclusion of the *Kyoto Protocol* treaty and the more recent *Paris Agreement* on climate change. It also birthed a worldwide movement for reshaping education that is now known as *Education for Sustainable Development (ESD)*. This effort gained support from the acceptance of the United Nations Decade of Education for Sustainable Development (2005–2014) by the United Nations General Assembly in 2002 and the UNESCO World Conference on Education for Sustainable Development in 2009 (Barth & Michelson, 2013). ESD is geared towards "enabling citizens to face the challenges of the present and future and leaders to make relevant decisions for a viable world" (UNESCO, 2005; p. 4). This broad goal is met through efforts to promote quality education; to reorient educational programs towards knowledge, skills, perspectives, and values related to sustainability; and to spread public awareness and professional development, such as with school teachers. With time ESD has acquired synonyms that stand for roughly similar efforts, such as *Education for Sustainability (EFS)* and *Learning for a Sustainable Future (LSF)*.

In a report summarizing the status of ESD in the United States, Smith, Rowe, and Vorva (2015) noted a diverse and prolific range of efforts under the overall ambit of USD that include programs in:

(i) Formal Education: in terms of professional development of K-12 educators, development of interdisciplinary curricula, and workforce training at the higher education level and youth engagement.
(ii) Non-formal Education: through grassroots-level efforts spearheaded by civil society associations, faith-based organizations, and public and private sector organizations.
(iii) Informal Education: by way of media and public awareness campaigns.

In the context of K-12 education, the focus in the United States has been most notable in the areas of professional development and the creation of "green schools" that have environment-friendly design and operations (Smith et al., 2015). Another area of progress is the increasing level of research in ESD as reflected in the birth of several academic journals in the last two decades that focus on issues related to ESD. The prominent examples being *Environmental Education Research, International Journal of Environmental and Science Education, Journal of Education for Sustainable Development,* and *Journal of Sustainability Education*. Despite these advances, however, it is hard to argue that ESD has led to any improvement in education in general and science education in particular in terms of better student understanding of our place in and relationship with the world (Frisk & Larson, 2011). Some researchers blame the lack of clear success on a lack of consensus on definitions, differences in understandings of aims and objectives, and diversity of practices (McFarlane & Ogazon, 2011). We, on the other hand, are skeptical of the potential for future ESD-related efforts on the grounds that they are based on a dated version of the school science curriculum that, as we have documented in this book, is inimical to the development of a scientific literacy needed for sustainable living in the

Anthropocene Epoch. What is needed, therefore, is an effort to bring school science into better alignment with the current state of scientists' science. We outline such an effort below.

Environmental Literacy Project at Michigan State University

Since 2003 a team of science educators at the Michigan State University, led by Charles W. Anderson, has been engaged in a pioneering effort to develop a research-based framework for environmental literacy for students from upper elementary school through college.[1] The group's work has been guided by the belief that "citizens must have an understanding of underlying scientific models and principles in order to evaluate experts' arguments about environmental issues and recognize policies and actions that are consistent with their environmental values" (Gunckel, Mohan, Covitt, & Anderson, 2012; p. 39). Underlying this project's efforts is a recognition that science education should reflect the emergence of environmental science as an interdisciplinary field, recognition of Earth's systems as coupled human and natural systems, and the emergent consensus on understanding environmental systems as dynamic and contingent (Anderson et al., 2004).

Consequently, the group conceptualizes environmental literacy as "the capacity to understand and participate in evidence-based discussions of socio-ecological systems and to make informed decisions about appropriate actions and policies" ("Environmental Literacy," n.d., para. 1). Over time, the environmental literacy framework developed by this team has taken shape in terms of developing learning progressions in three strands that represent critical areas of school science curricula and environmental literacy:

Carbon: This strand focuses on "Carbon-transforming processes in socio-ecological systems at multiple scales, including cellular and organismal metabolism, ecosystem energetics and carbon cycling, carbon sequestration, and combustion of fossil fuels." ("Environmental Literacy," n.d., para. 3; Mohan, Chen, & Anderson, 2009).

Water: This strand covers topics that relate to "The role of water and substances carried by water in earth, living, and engineered systems, including the atmosphere, surface water and ice, ground water, human water systems, and water in living systems." ("Environmental Literacy", n.d., para. 4; Gunckel et al., 2012).

Biodiversity: Here the emphasis is on "The diversity of living systems, including variability among individuals in population, evolutionary changes in populations, diversity in natural ecosystems and in human systems that produce food, fiber, and wood." ("Environmental Literacy," n.d., para. 5; Hartley et al., 2011).

These learning progressions are designed to work as conceptual frameworks for development of science curricula, assessment, and teaching practices that help students progress towards appropriation of science as a secondary discourse such that they begin to use model-based reasoning to understand the world around them.

The Environment Literacy Project is a good representation of current work being done in the United States that aims to re-conceptualize science education for developing environmental science literacy. Much of this work is currently focused on including climate change as an important topic in science education (see for example Shepardson, Roychoudhury, & Hirsch, 2017). This book supports and builds upon such work. Indeed, the research and arguments offered here were inspired by the pathbreaking work done by the Michigan State University project. We intend this book to serve as the groundwork for the next stage of re-envisioning science education for the Anthropocene Epoch. In the next section, we outline the key elements of our proposal for this important next step.

SCIENCE EDUCATION FOR THE AGE OF WICKED PROBLEMS: OUR PROPOSAL

Imagine that a student learns in a science classroom one day that bananas, the most consumed fruit in the United States, comes from plantations that have caused massive destruction of rainforests in South and Central America (Clay, 2013). This student may decide that a boycott of bananas would be the best way to save rainforests from these plantations. In fact, many mainstream environmental groups, such as *Rainforest Relief*, do urge customers to "avoid purchasing bananas altogether and instead opt for fruit grown locally, such as apples, peaches, cherries or pears" ("Banana industry's impact on Rainforests", 2010). Alternately, some environmental groups, *Rainforest Trust*, for example, may try to save rainforests by buying land in these regions so that they can be restored to their pristine ecological health (Butler, 2014). But as Vandermeer and Perfecto (2005) explain, such actions alone may hurt the rainforests more than save them. The closure of banana plantations can result in laying off of plantation workers who often end up converting forests into subsistence farmlands in order to survive.

This student may instead decide that buying organic bananas might be the best option to help save the rain forests. However, the world currently is not in a position to feed all the people on the planet through organic farming (Seufert, Ramankutty, & Foley, 2012). Organic bananas can be grown in only very specific conditions that severely limit the amount of land available for growing them. So even if there was a 10% percent drop in supply of regular bananas, the potential of growing organic bananas will not be able to meet the demand (Loza, 2016). The cost of production for organic bananas is much higher too. So, if only organically grown bananas were available in the grocery stores, it could mean that bananas would go back to being the exotic fruit for the rich like they were back in the nineteenth century. Again, higher prices may decrease demand, laying off plantation workers who return to unsustainable subsistence farming practices. Similar outcomes may result if our student adopts the strategy of raising money to buy up land for conservation and restoration. This is not likely to work either and may only lead to an ecological landscape marked by "isolated islands of tropical rain forest surrounded by a sea of pesticide-drenched modern agriculture, underpaid rural workers, and masses of landless peasants looking for some way to support their families" (Vandermeer & Perfecto, 2005; p. 13).

A seemingly simple question of whether to consume or boycott bananas ends up revealing a complex global assemblage of relations and entanglements involving local and distant human, nonhuman, material, social, and cultural actors and ethical-political dimensions. Simple actions such as a product boycott can indeed be counterproductive in resolving environmental issues because when we effect one strand of the complex web of causality inherent in these assemblages, the effects reverberate through the web in unanticipated ways to yield all kinds of desirable and undesirable outcomes. Another example of simple environmentally minded actions leading to negative and unanticipated consequences would be the increasing of biofuel production in recent years to reduce the emission of climate change-causing gases. This single policy move has led to direct and indirect land-use changes in ways that have put food security, biodiversity, and local livelihoods at risk (Harvey & Pilgrim, 2011; Tilman et al., 2009). Unsurprisingly, therefore, environmental problems are often seen as "wicked problems" that can't be solved by either science or by social action alone (Brown, Harris, & Russell, 2010).

Developing on ideas that Horst Rittel conceptualized in the mid-1960s, Horst Rittel and Melvin Webber (1973) wrote an iconic paper that first formalized a theorized notion of "wicked problems" in areas of social policy. According to them, a scientific approach is bound to fail in solving problems of social policy because "They are "wicked" problems, whereas science has developed to deal with "tame" problems" (p. 155). They then listed ten key features of societal problems that make them "wicked," such as there being no definitive way to conceptualize a wicked problem and such problems having no exhaustively describable set of solutions. In the years since, researchers have come to recognize environmental issues as a classic example of wicked problems (Camillus, 2008). Environmental problems are wicked because they are "defined by high complexity, uncertainty, and contested social values" (Miller, 2013, p. 279). They arise from "the functioning and evolution of interconnected and complexly interacting socio-ecological systems" and defy solutions because "they are multicausal, intertwined with other problems, and value-laden" (Metzger & Curren, 2017; p. 94). We suspect that the increasing proliferation of "quasi-objects" or "naturecultures" in our world, such as climate change, genetically modified or cloned animals, and the ozone hole, each of which emerges out of a complex mixture of natural, social, cultural, and political entities and relations, may have made environmental issues progressively more wicked with each passing decade.

Wicked problems define our existence in the Anthropocene Epoch. Therefore, if we wish to remain hopeful about our future, we need to prepare our students as citizens who are able to appreciate the "wicked" nature of environmental issues facing our planet and understand the type of science that is needed to find ethical and socially just solutions for such problems. Wicked problems can rarely be addressed by conceptual and methodological tools and resources available in any single discipline. Therefore, it is not surprising that sustainability science is increasingly being seen as the preferred approach to understanding and resolving the wicked environmental issues of our world (Hart & Bell, 2013). This is because sustainability science is a transdisciplinary field of research "dealing with the interactions between natural and social systems, and with how those interactions affect the challenge of sustainability: meeting the needs of present and future generations while substantially reducing poverty and conserving the planet's life support systems" ("Sustainability Science", n.d.).

Therefore, based on the three core research dimensions of sustainability science as summarized in Dedeurwaerdere (2013), we propose a sustainability science-based framework for science education that is (a) imbued with an **ethical stance** that celebrates equity, democratic contestation, reconciliation through negotiation, and a virtue-based ethics of care, kindness, and compassion; (b) compatible in terms of **content** with the latest developments in ecology and environmental sciences; and (c) oriented towards **praxis** in the service of social-ecological justice.

As can be seen in Fig. 7.1, our framework posits ethics as the foundation of teaching and learning about the world. This is because following Levinas (1979), we see ethics as the "first philosophy" that precedes any effort to understand the world. As Levinas has said, "there is a vigilance before any awakening that the *cogito* (emphasis in original) is possible, in such a way

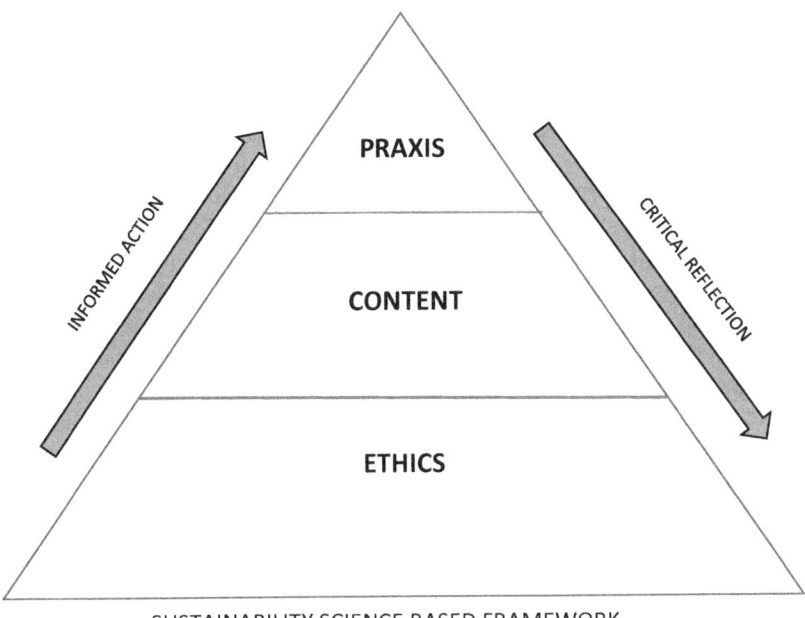

Fig. 7.1 Sustainability science-based framework for science education

that ethics is before ontology" (Levinas & Robbins, 2001; p. 211). Thus, any relation with the other is primarily an ethical relation that obligates us into a relationship of responsibility with the rest of the world. We would want our students to appreciate the inevitable ethical nature of their presence in, and interactions with, the world before and as they come to know about the world through school science, in particular, and school-based education in general. Our decision to start with ethics can also be seen as a corrective response to the problems inherent in the dominant educational paradigm that embraces an instrumental rationality that erases ethics from knowledge about the world (Giroux, 2011). Thus, realizing teaching and learning as primarily an ethical relation, and education as a values-based enterprise, we support a sustainability science-based science education that promotes an ethical commitment to intellectual rigor, civility, and care (Christie, 2005).

Viewing science education from the perspective of Aristotelian categorization of knowledge (Welldon, 1897), it can be said that by occluding ethical considerations from science content, science education in our modern times has primarily focused on enabling students to acquire *epesteme* (scientific knowledge) and *techne* (technical knowledge). Thus, by beginning with ethical considerations, students' understanding of science content will come to have an additional much-needed dimension—that of *phronesis*, which can be roughly translated as prudence or practical wisdom. As Flyvbjerg (2001) asserts "*Phronesis* is most important because it is that activity by which instrumental rationality is balanced by value-rationality" (p. 4), so that students do not only come to acquire knowledge in science classrooms but also the wisdom to use that knowledge for the public good. Of late, science education in the United States has been moving from a focus on science content to a greater emphasis on student appropriation of scientific practices and discourses that enables them to participate in society as educated citizens (Duschl, Schweingruber, & Shouse, 2007). As we noted in the third chapter, this emphasis on scientific practices is now well reflected in the national- and state-level science standards. We support this change but would urge the science education community to go much further to embrace a critical pedagogical orientation towards praxis as a goal in science education. That is, we believe that an ethics-based understanding of science content should lead students to praxis, understood from a critical pedagogical standpoint as informed action that results from a complex combination of theory and practice (Kincheloe, 2008). Thus, we support Roth and Desautel's (2002) appeal for science education as/for sociopolitical action when they state that rather than thinking of school science in terms of scientists' science, we may think of it in terms of participation in public affairs related to science and technology policy. Furthermore, this participation does not have to be delayed to some future point in time. Rather ... children and young adults are perfectly able to participate in science-related activities that contribute to a larger good. (p. 7)

Further, as we indicate in Fig. 7.1, just as knowledge undergirded by ethical commitments of intellectual rigor, civility, and care leads to informed action, engagement in praxis should occasion critical reflections back upon science content and its ethical considerations.

We believe that science education based on such a framework will orient students to see the world in terms of a relational agent-based ontology that gives nonhumans as much agency as humans and accords primacy to relations over entities in descriptions of the world. In this way, it will eschew the dualism between anthropomorphism and ecocentrism that typifies our current normative constructions of nature. This framework, instead, adopts a noncentered democratic ecologism that insists on problematizing "nature" and "social" as sociopolitically and discursively contingent ontological categories and thus understands "'nature' as multiform and as inextricably confounded with humanity's projects and selfunderstandings" (Whiteside, 2002; p. 3). In the remainder of the section, we elaborate on the three key facets—ethics, content, and praxis—of our framework.

Ethics

Schools science curricula, as currently written and enacted, expect students to grow up as environmentally responsible citizens. However, the ethical stance offered to them is marked by three key features that orient them towards practices that have troublesome ethical implications. First, this ethical stance is based on a strong belief in human exceptionalism that leads students to partition the world in two distinct social and natural realms. This belief positions human concerns at the center and issues related to nonhuman existence and survival at the periphery. Second, ethical decisions are expected to be taken on the basis of instrumental reasoning that, in concert with human exceptionalism, supports the commodification of the nonhuman aspects of our world. Third, these curricula support a neoliberal ethic in which our daily life choices, in terms of consumption and lifestyles, are seen as matters of individual choice and environmental action as personal virtue. As a result, students may grow up as pro-environment citizens who, despite their reuse, recycle, and reduce commitments, remain bound to an environmentally destructive and socially unjust political-economic regime because they lack the tools and orientation to use collective action to force systemic progressive change.

Science curriculum, in its bid to be objective, shies away from acknowledging its ethical orientations. This, in our view, is neither successful nor advisable. Understanding and acting on the world from a scientific perspective comes with momentous ethical implications that need to transition from the implicit to explicit domains of the science curricula if we wish to educate students as reflective and reflexive ethical actors. Our sustainability science framework for science education explicitly acknowledges an ethical stance focused on noncentered democratic ecologism that does away with a scientifically objectified and commodified "nature" and the consequent sundering of the "social" from the "natural." As a result, we wish to see students receiving a critical introduction to a relational ethical standpoint while learning about ecological issues and topics. This ethical stance, we believe, should be as much of an object of critical scientific inquiry as other components of the science curricula.

A noncentered democratic ecologism favors following ethical guidelines. First, it explicitly recognizes "webs of connectivity between the livelihood practices and cultural values of particularly situated human actors (collective and individual) and the life-habits and relationships of other biotic agents" in order to support a virtue-based ethics of care, kindness, and compassion "both in the sense of caring about 'generalized others' and caring for 'concrete others'" (Whatmore, 1997; p. 45). Second, by discarding the nature-social dualism, noncentered democratic ecologism encourages us to see the world as consisting of networks of nature-culture collectives (Latour, 2012). In these nature-culture collectives, nonhumans are no longer relegated as objects with no ethical standing. Instead, they get included as constituent members of the social with the understanding that we extend equivalent (if not equal) ethical obligations to them as accorded to the humans. Of course, scholars are still in the early stages of figuring out how to include the nonhuman world in our social justice considerations. Summarizing work done in this direction, Roe (2016) suggests that we can adopt Joana Formosinho's approach "to use human imagination to explore what it is like to be a plant, an animal, a river, through the use of thick description derived from scientific knowledges, lay knowledges and creative thought, to encourage humans to explore becoming cow-ness, or becoming river-ness as an embodied experiment of becoming" (p. 1953).

Alternately, we can follow Michelle Bastian and her colleagues in efforts "to try and become 'in conversation' with non-humans," such as by asking "how the non-human world can be included within our ethical protocols" (p. 1954). It is our hope that when we adopt such perspectives on understanding the nonhuman world, we might become more responsive to the nonhuman world, and thus do a far better job at minimizing harm and suffering to the nonhumans we share this planet with.

Third, because we are ourselves embedded in networks of nature-culture collectives, noncentered democratic ecologism invites us to liberate ourselves from "the geographical imaginary of ethical community from the territorialized spaces of the embodied individual, the local neighborhood, and the nation-state" so as to extend our ethical obligations to all distant and proximate members of our networks (Whatmore, 1997). Thus, for instance, when we buy and consume bananas from a grocery store we can learn to see ourselves as ethically connected and obligated to all the human and nonhuman actors in the hybrid networks or global assemblages devoted to banana production and consumption regardless of their spatiotemporal proximity to us. Finally, noncentered democratic ecologism commits us to ethical pluralism. That is, it does not privilege any particular set of values in our relations with other human and nonhuman members of our nature-culture collective networks. As a result, it threatens and delegitimizes the hegemony of neoliberal ethics. Instead on each issue it encourages us to democratically participate in "a deliberative process in which every constituent of a hybrid-forming network gets a chance to challenge others' views of the facts and to probe the appropriateness of their research methods, the reliability of their measuring instruments, and the moral acceptability of their political and economic connections" (Whiteside, 2002; p. 279). Concomitantly, ethical environmental action becomes less a reflection of our personal virtue and more an outcome of our commitment to democratic collective action.

Content

In alignment with the foundational assumptions of the Environmental Literacy Project at Michigan State University (http://envlit.educ.msu.edu/), we wish to see a K-12 science education that prepares students to understand our world as consisting of continually evolving complex, adaptive, resilient, and integrated socioecological systems. In ecological sciences, there are currently at least ten established frameworks for analyzing social-ecological systems. As Binder, Hinkel, Bots, and Pahl-Wostl (2013) report, these frameworks "differ significantly with respect to contextual and structural criteria, such as conceptualization of the ecological and social systems and their interrelation" as they have been designed to fulfill different needs and theoretical commitments. In addition, within the science education community there are two major models currently in circulation. One of them is the "loop diagram" model of social-ecological systems proposed by the environmental literacy project at the Michigan State University. This model conceptualizes these systems as consisting of two ontologically distinct "Human Social and Economic Systems" and "Environmental Systems" that are coupled by matter and energy transferring relationships of waste disposal and ecosystem services consumption (Gunckel et al., 2012). The second model proposed by Ruppert and Duncan (2017) is a revised version of the Michigan State model. Ruppert and Duncan's model shows humans as embedded within environmental systems at various scales. It also shows flows of ecosystem services and natural resources between human populations and environmental systems, and also among human populations.

Rather than endorsing one particular model of social-ecological systems for our framework, we are content for now in encouraging readers to explore all of these alternatives to determine which one would work best for their pedagogical and research purposes. This is still a new field that needs more research and curriculum development work. We are clear, however, that any chosen model should encourage the students to understand a social-ecological system as a complex system in which the "social" and "ecological" are not two distinct though coupled subsystems, but rather, each is a component of a unified system with two interacting social and ecological dimensions. More specifically, we expect that by the end of their high school education, students should understand that such complex systems are defined by the following key attributes:

(i) *Nonlinearity*: Rather than following linear energy flows and matter cycles, complex socioecological systems are defined by nonlinear dynamics in which feedback loops, time lags, and other complex behaviors lend an element of unpredictability to the outcomes (Berkes, Charles, & Loucks, 2014). Scientists have tried to model such systems through nonlinear mathematical equations. An interesting property of nonlinear mathematical models is that "mathematical solutions to nonlinear equations do not give simple numerical answers but instead produce a large collection of values for the variables that satisfy an equation" (Berkes, Colding, & Folke, 2003; p. 5). Thus, complex systems can have several possible equilibrium states. This mathematical property combined with the fact that the social valuation of each possible value or equilibrium state will be discursively, sociopolitically, spatially, and historically contingent make nonlinear socioecological systems notorious for resisting solutions that are optimal or designed to maximize efficiency from an economic perspective. In fact, research shows that "the more you optimize elements of a complex system of humans and nature for some specific goal, the more you diminish that system's resilience. A drive for an efficient optimal state outcome has the effect of making the total system more vulnerable to shocks and disturbances" (Walker, Salt, & Reid, 2012; p. 9).

(ii) *Scale*: Socioecological systems often comprise hierarchically organized, nested subsystems at each level of organization. For instance, many *patches* may combine to constitute an *ecosystem* which then spatially link up with other ecosystems to form a *landscape* in a multiscale topological ecology. Similarly, "institutions may be considered hierarchically, as a nested set of systems from the local level, through regional and national, to the international" (Walker et al., 2012; p. 6). Further, such systems are marked by cross-scale interactions and feedback relations that have ecological and social dimensions through both space and time. These interactions and relations give rise to emergent properties at each level. For instance, as Simon Levin (2005) explains, "The patterns that characterize ecosystems—the distribution and abundance of species, and their spatial organization, size structure distributions, and patterns of nutrient use (stoichiometry)—all can be realized as emergent from selection forces operating at much lower levels of organization" (p. 1077). Thus, as Walker et al. (2012) opine, "complex systems should be analyzed or managed simultaneously at different scales" (p. 6).

(iii) *Self-organization*: Studies show that in a complex system multifaceted patterns of interactions and structures emerge from disorder and chaos through successive iterations of a few simple rules that guide change (Folke, 2006). Self-organization has been found to be a defining characteristic of complex socioecological systems.

To give a simple and common example, it has been seen that all leaves in a deciduous boreal forest turn to face the sun to maximize the uptake of solar energy (Olsson, Jerneck, Thoren, Persson, & O'Byrne, 2015). Further, if disturbed at a critical point of instability, socioecological systems are able to reorganize into a new metastable equilibrium state through feedback mechanisms (Walker, Salt, & Reid, 2012). For an example, readers may refer to Filotas et al.'s (2014) analysis of illegal logging as a self-organizing phenomenon in Borneo. Their analysis shows how illegal logging and the resulting destruction of the forest emerge as a meta-stable equilibrium state arising from interactions between social-ecological components at all levels in that forest ecosystem.

(iv) *Ecological flows*: At present, following a cybernetic model, students in their science classes learn to trace matter cycles and energy flows within an ecosystem. This is because school science follows a traditional ecosystem ecological perspective that has always typically emphasized "understanding changes through time within a single ecosystem rather than understanding variation across space" (Turner & Cardille, 2007; p. 62). However, recent developments in ecology suggest that real world ecosystems are never closed systems, and the traditionally ignored "noise" originating from spatial heterogeneity is actually critical for understanding ecosystem processes. That is, we find that even when globe spanning anthropogenic flows are not factored in, matter and energy transfer across ecosystem boundaries, rather than being an exception, are the norm for the world we see around us. As Polis, Anderson, and Holt (1997) assert, "Movements of nutrients, detritus, prey, and consumers among habitats are ubiquitous in diverse biomes and can strongly influence population, consumer-resource, food web, and community dynamics" (p. 289). For instance, in a study of a Sonoran Desert stream, Jackson and Fisher (1986) found that more than 96% of the insect biomass was transferred to the adjacent terrestrial ecosystem. Similarly, ecologists now recognize the importance of herbivores in redistributing nutrients across landscapes (Turner & Cardille, 2007). Further, it is not only matter and energy that flow across ecosystem boundaries. Ecologists now recognize four kinds of ecological flows that are critical for understanding social-ecological systems (Cadenasso, Pickett, Weathers, & Jones, 2003; Jianguo, 2013). These are:

(a) Material flows: Matter cycling is a core concept in school science. However, in our view we need to expand this topic to include matter flows across ecosystem boundaries. As discussed above, material, such as detritus, nutrients, and pollutants, flows across ecological boundaries in ways that vary both spatially and temporally in terms of type, amount, and delivery mechanism, and this spatial variation is critical to understanding social-ecological systems. Such flows have become increasingly important in understanding the fate of our social-ecological systems because of the ever-expanding global reach of industrialized production and consumption systems of our world. For instance, one cannot have a realistic understanding of nitrogen cycles in today's world without factoring in the role that global agro-food system has played in profoundly transforming the nitrogen cycle at all levels of social-ecological systems throughout the world through mass production and distribution of chemical fertilizers (Mancus, 2007).

(b) Energy flows: Similarly, energy, stored in biological forms as well as carried through vectors like wind and tides, flows not just within an ecosystem but across all boundaries and scales in social-ecological systems. For example, "material consumed by an animal in one patch and defecated in a different patch may transport stored energy from the first patch for use by decomposers in the second patch" (Cadenasso et al., 2003; p. 752). Industrialized production and consumption systems in our world are indeed premised on sustained energy flows across social-ecological systems. Aquaculture would be a good example as studies show that farming of fish and other aquatic creatures, especially in industrialized nations, is critically dependent upon "the redirection, concentration and dissipation of various forms of energy from the environment" (Tyedmers & Pelletier, 2007; p. 231).

(c) Organism flows: Annual migration of humpback whales between the icy and warm Hawaii waters and of songbirds along the entire length of the North American continent are well-recognized annual events. Organisms flow occurs at smaller distances as well, such as through seed dispersal in case of plants or movement of microbes between soil horizons (Cadenasso et al., 2003). Ecologists now recognize that most social-ecological systems cannot be well understood without including the flow of organisms across social-ecological scales and boundaries, especially when most organism flow regimes have been deeply impacted by human activity. Famous cases include the severe threat to migration and hence survival of red knot shorebirds because of harvesting of horseshoe crabs along the Delaware coast for medicinal purposes (Cramer, 2015). Each year red knots undertake a 19,000 mile journey along the length of the two American continents and critically depend upon horseshoe crab eggs for nutrition to be able to make this trip.

(d) Information flows: Along with matter, energy, and organisms, another kind of flow is now recognized as important for its role in social-ecological phenomena and processes. This is the flow of information through social-ecological boundaries (Manel, Schwartz, Luikart, & Taberlet, 2003), which can be in the form of flow of genes as well as transfer of visual, auditory, and chemical signals between spatially distributed organisms (Cadenasso et al., 2003). For instance, while gene flow occurs naturally and is critical to the resilience of social-ecological systems, studies have also been able to document the profound impact on domesticated and wild flora on account of commercial, state-regulated, as well as informal human-mediated germplasm flow in plant species (Ellen & Platten, 2011; Warwick, Beckie, & Hall, 2009). Ecologists also study the role of information flows along the social networks in the governance of social-ecological systems (Spirn, 2014).

(v) *Social-ecological resilience*: Complex social-ecological systems manifest resilience—an attribute that allows these systems to preserve themselves, especially in terms of structure and function, despite disruptions and change (Berkes et al., 2003). Systems with high resilience are able to "tolerate disturbance events without collapsing into a qualitatively different state that is controlled by a different set of processes" ("Resilience"; n. d.). Humans naturally play a big role in shaping resilience of social-ecological systems. They not only influence systems' future pathways but also are critical for the ability of social-ecological systems to anticipate and adapt to change. Resilience-oriented research has been instrumental in helping ecologists and environmental scientists appreciate that stability and resilience are not synonymous when it comes to social-ecological systems. In fact, attempts to increase the stability of social-ecological systems can even reduce their resilience (Holling, 1973). As a result, stability and equilibrium is no longer seen as a mark of good health for social-ecological systems.

Further, resilience-based research has enabled ecologists to discard "the classic model of ecological succession in natural communities in which the ecosystem progresses toward a steady-state, climax condition as it changes" in favor of a resilience-based framework of adaptive cycles in social-ecological systems in which "succession is only a part of the cycle that also includes the distinct yet interconnected phases of exploitation, conservation, destruction leading to decline, and finally renewal" (Curtin & Parker, 2014; p. 915). Students in schools still learn about change in social-ecological systems in terms of linear ecological succession. We see no reason why students cannot learn about change through adaptive cycles instead. Students should also learn that the resilience of social-ecological systems is limited by thresholds beyond which if a system is perturbed it loses enough of its resilience to reorganize itself into a new regime with a different set of processes and structures (Garmestani & Benson, 2013). The issue of thresholds is critical for governance of social-ecological systems and has acquired increasing importance because ecologists now fear that "the future of human well-being may be seriously compromised if we should pass a critical threshold that tips the earth system out of this stability domain" (Folke et al., 2010; p. 2).

(vi) *Environmental governance*: Whether we like it or not, there is now no patch of land or water on the surface of this planet that has not been impacted by human activity directly or indirectly. Thus, rather than advocating a hands-off approach to rejuvenation of Earth's social-ecological systems, we are of the view that governance of social-ecological systems needs to be seen as, and in fact, taught to students as an intrinsic feature of all social-ecological systems. Further, students should be able to connect the health of a social-ecological system with its governance. This is because rather than seeing "people," as responsible for the current state of the planet, it is important that students realize that "the Anthropocene is an epoch characterized not only by the anthropogenic dominance of the Earth's ecosystems but also by new forms of environmental governance and institutions ... we call these new forms of environmental governance 'global assemblages'" (Ogden et al., 2013; p. 341). These global assemblages usually comprise powerful transnational corporations, organizations, and state governments. Protected by the false assurances of green capitalism and neoliberal discourse, global assemblages have long proven to be profoundly undemocratic, unsustainably exploitative, and disastrous for communities on the margins as well as nonhuman forms of life (Büscher et al., 2014; Kraemer, 2012).

Given such a scenario, it should be very useful for students as future agentive citizens to understand that new models of collaborative, nonexploitative governance, collectively known as adaptive governance systems, have also become increasingly important in the last few decades. According to Folke, Hahn, Olsson, and Norberg (2005), "Adaptive governance systems often self-organize as social networks with teams and actor groups that draw on various knowledge systems and experiences for the development of a common understanding and policies" for long-term coordinated action (p. 441). Examples of these governance systems include Landscape Conservation Cooperatives (LCC) which are networks of governmental and nongovernmental conservation organizations. There are 22 LCCs in the United States. Each LCC "brings together federal, state, and local governments along with Tribes and First Nations, non-governmental organizations, universities, and interested public and private organizations" to form collaborative partnerships that "leverage resources, share scientific expertise, fill needed science gaps, identify best practices, and prevent duplication of efforts through coordinated conservation planning and design" ("Landscape Conservation Network"; n.d.). Here we should also take note of transnational networks of grassroots organizations that in recent decades have played a critical role in environmental governance by supporting indigenous populations all over the world in their efforts to stave off ecological depredation of their native lands from local actors and transnational corporations (Ogden et al., 2013).

Praxis

Following Aronowitz and Giroux's (1985) advocacy of critical pedagogy in public education, our proposal positions science teachers as transformative intellectuals who will extend the boundaries of learning beyond textbooks and classrooms into the society at large. By orienting students to critically and reflexively deliberate on their ethical commitments on issues related to our relations with the nonhuman world, and by equipping them with the most up-to-date scientific understanding of the world, such teachers are expected to transform students into critical thinkers and agents of positive social-ecological change in whatever networks of nature-culture collectives they are embedded in. In this way, as readers will notice, our framework aligns well with the socioecojustice current in the overall Science-Technology-Society Education (STSE) movement in science education. That is, in line with key tenets of socioecojustice science education, we also believe that "the aim of science education should be the promotion of a certain type of citizenship and civic responsibility of which transformation, agency, and emancipation are key features" (Pedretti & Nazir, 2011; p. 617). Further, our framework also shares many of the goals of citizen science, especially in terms of its commitment towards participatory democracy and civic responsibility and ethical obligations to all human and nonhumans on this planet (Mueller & Tippins, 2012).

As should be clear from our earlier arguments, when we see students acting as agents of change we are not primarily thinking of individual actions, such as writing letters to members of the government and business leaders or educating others on STSE issues. Such actions are indeed important in STSE education, especially in terms of inculcating a performativity of environmental action in the beginning stages. However, as we have pointed out, such actions do not match the ecological and sociological scales at which socioecological problems express themselves, and on their own, do not lead to much change. Thus, starting from simpler individual-based actions, we would like to see students maturing into engaged actors in collective social-ecological justice struggles for both humans and nonhumans.

Looking Ahead

Our critique of existing written and enacted science curricula was an attempt to disarticulate some powerful "official" representations of our social-ecological world that have become naturalized in science education to the point that it is often difficult to imagine alternative representations. In this work, we benefitted enormously from the existing critiques of science curricula by a small group of science education researchers such as Tasos Hovardas, Konstantinos Korfiatis, Elizabeth Hufnagel, John Ruppert, and Noah

Feinstein. However, if we wish to see change in this world, deconstruction alone, albeit necessary, is not enough. It needs to be followed by reconstruction as well (Fraser, 1995). That is, when established dominant interpretations are deconstructed and destabilized, the resulting creative space needs to be filled with fresh constructive projects that offer better alternatives for our future. Building on pioneering work done by the Environmental Science Literacy Project at the Michigan State University, we have proposed a framework for science education on ecological topics that goes beyond content to include matters of ethics and praxis.

Admittedly, our proposal is just a first (and small) step that needs to be strengthened with further research and curriculum development efforts. Nonetheless, we firmly believe that the overall direction for future work as indicated in our proposal is not much off the mark. We live in a complex world filled with wicked problems where the old-fashioned reductionist science with its false ideology of human mastery of all things in this world is proving to be of little help (Laughlin, 2008). Teaching this view of science to students does little to prepare them for the world that confronts them with wicked problems. Tomorrow's decision-makers will need a view of science that is transdisciplinary in intent as well as content. In recent years, science education has indeed become more transdisciplinary as it is increasingly yoked to engineering and technology education through STEM education reforms. This is, however, a step in exactly the opposite direction of the changes we are advocating, in that STEM education reform is intended to serve the interests of the US economy and not it's (and the world's) ecology (Sharma, 2016). It is our hope that science teachers, teacher educators, researchers, and policymakers are able to come together to initiate reforms that orient school science more towards helping students create a sustainable democratic and socioecologically just world, rather than serving the material interests of the moneyed elite. For this to occur, we believe that rather than aligning science more closely with engineering, we would need to bring science closer to social studies, so that students can be better prepared to understand and act on social-ecological wicked problems that afflict our planet. A sustainability science-oriented science education definitely opens possibilities for conceptualizing a Science-Social Studies (S^3) education movement in our schools. We hope that this book contributes to such ends.

Note

1. One of us (Sharma) was a member of the project in its initial phase from 2003 to 2006.

REFERENCES

Anderson, C. W., Sharma, A., Lockhart, J., Carolan, A., Moore, F., Parshall, T., & Gallagher, J. (2004). *Partial draft: Environmental literacy blueprint.* East Lansing, MI: Michigan State University.

Anderson, M. (2017). For Earth Day, here's how Americans view environmental issues. Pew Research Center. Retrieved from http://www.pewresearch.org/facttank/2017/04/20/for-earth-day-heres-how-americans-view-environmental-issues/

Aronowitz, S., & Giroux, H. A. (1985). Teaching and the role of the transformative individual. In *Education under seige: The conservative, liberal and radical debate over schooling* (pp. 23–45). South Hadley, MA: Bergin & Garvey.

Assaraf, O. B.-Z., & Damri, S. (2009). University science graduates' environmental perceptions regarding industry. *Journal of Science Education and Technology, 18*(5), 367–381.

Banana industry's impact on Rainforests. (2010). *Business Ethics.* Retrieved from http://business-ethics.com/2010/06/19/2440-banana-industrys-impact-on-rainforests/

Barth, M., & Michelsen, G. (2013). Learning for change: An educational contribution to sustainability science. *Sustainability Science, 8*(1), 103–119.

Berkes, F., Charles, A., & Loucks, L. (2014). *Guidelines for analysis of social-ecological systems.* Retrieved from http://www.communityconservation.net/wp-content/uploads/2016/01/FINAL_CCRN-Guidelines-for-Analysis-of-Social-Ecological-Systems-September-2014.pdf

Berkes, F., Colding, J., & Folke, C. (2003). *Navigating social-ecological systems: Building resilience for complexity and change.* Cambridge, UK: Cambridge University Press.

Binder, C., Hinkel, J., Bots, P., & Pahl-Wostl, C. (2013). Comparison of frameworks for analyzing social-ecological systems. *Ecology and Society, 18*(4).

Brown, V. A., Harris, J. A., & Russell, J. Y. (2010). *Tackling wicked problems through the transdisciplinary imagination.* Washington, DC: Earthscan.

Büscher, B., Dressler, W., & Fletcher, R. (2014). *Nature Inc.: Environmental conservation in the neoliberal age.* Tucson, AZ: University of Arizona Press.

Butler, R. (2014). Saving rainforests by buying them. *Mongabay.* Retrieved from Mongabay website: https://news.mongabay.com/2014/04/saving-rainforests-by-buying-them/

Cadenasso, M. L., Pickett, S. T., Weathers, K. C., & Jones, C. G. (2003). A framework for a theory of ecological boundaries. *AIBS Bulletin, 53*(8), 750–758.

Camillus, J. C. (2008). Strategy as a wicked problem. *Harvard Business Review, 86*(5), 98.

Christie, P. (2005). Towards an ethics of engagement in education in global times. *Australian Journal of Education, 49*(3), 238–250.

Clay, J. (2013). *World agriculture and the environment: A commodity-by-commodity guide to impacts and practices*. Washington, DC: Island Press.

Cramer, D. (2015). *The narrow edge: A tiny bird, an ancient crab, and an epic journey*. New Haven, CT: Yale University Press.

Curtin, C. G., & Parker, J. P. (2014). Foundations of resilience thinking. *Conservation Biology, 28*(4), 912–923.

Davenport, C., & Nagourney, A. (2017). Fighting Trump on climate, California becomes a global force. *The New York Times*. Retrieved from https://www.nytimes.com/2017/05/23/us/california-engages-world-and-fights-washington-on-climate-change.html

Dedeurwaerdere, T. (2013). *Sustainability science for strong sustainability*. Report on the Organisation of Scientific Research, Universite catholique de Louvain, Louvain-la-Neuve.

Driver, R., Asoko, H., Leach, J., Mortimer, E., & Scott, P. (1994). Constructing scientific knowledge in the classroom. *Educational Researcher, 23*(7), 5–12.

Dryzek, J. S. (2013). *The politics of the earth: Environmental discourses*. Oxford, UK: Oxford University Press.

Duschl, R. A., Schweingruber, H. A., & Shouse, A. W. (2007). *Taking science to school: Learning and teaching science in grades K-8*. Retrieved from Washington, DC: National Academies Press.

Ellen, R. O. Y., & Platten, S. (2011). The social life of seeds: The role of networks of relationships in the dispersal and cultural selection of plant germplasm. *Journal of the Royal Anthropological Institute, 17*(3), 563–584. https://doi.org/10.1111/j.1467-9655.2011.01707.x

Environmental literacy. (n.d.). Retrieved from http://envlit.educ.msu.edu/

Fairclough, N. (2003). *Analysing discourse: Text analysis for social research*. London, UK: Routledge.

Feinstein, N. W., & Kirchgasler, K. L. (2015). Sustainability in science education? How the next generation science standards approach sustainability, and why it matters. *Science Education, 99*(1), 121–144. https://doi.org/10.1002/sce.21137

Fendler, L. (2008). New and improved educationalising: Faster, more powerful and longer lasting. *Ethics and Education, 3*(1), 15–26.

Filotas, E., Parrott, L., Burton, P. J., Chazdon, R. L., Coates, K. D., Coll, L., ... Messier, C. (2014). Viewing forests through the lens of complex systems science. *Ecosphere, 5*(1), 1–23. https://doi.org/10.1890/ES13-00182.1

Flyvbjerg, B. (2001). *Making social science matter* (trans: Sampson, S.). Cambridge, UK: Cambridge University Press.

Folke, C. (2006). Resilience: The emergence of a perspective for social–ecological systems analyses. *Global Environmental Change, 16*(3), 253–267.

Folke, C., Carpenter, S., Walker, B., Scheffer, M., Chapin, T., & Rockström, J. (2010). Resilience thinking: Integrating resilience, adaptability and transformability. *Ecology and Society, 15*(4).

Folke, C., Hahn, T., Olsson, P., & Norberg, J. (2005). Adaptive governance of social-ecological systems. *Annual Review of Environment and Resources, 30*(1), 441–473. https://doi.org/10.1146/annurev.energy.30.050504.144511

Fraser, N. (1995). Pragmatism, feminism, and the linguistic turn. In S. Benhabib (Ed.), *Feminist contentions: A philosophical exchange* (pp. 157–171). London, UK: Routledge.

Frisk, E., & Larson, K. L. (2011). Educating for sustainability: Competencies and practices for transformative action. *Journal of Sustainability Education, 2*(1), 1–20.

Gardner, H. (2011). *The unschooled mind: How children think and how schools should teach.* New York, NY: Basic Books.

Garmestani, A. S., & Benson, M. H. (2013). A framework for resilience-based governance of social-ecological systems. *Ecology and Society, 18*(1), 9.

Giroux, H. A. (2011). *On critical pedagogy.* New York, NY: Bloomsbury Publishing.

Gunckel, K. L., Mohan, L., Covitt, B. A., & Anderson, C. W. (2012). Addressing challenges in developing learning progressions for environmental science literacy. In A. C. Alonzo & A. W. Gotwals (Eds.), *Learning progressions in science* (pp. 39–75). Rotterdam, The Netherlands: Sense Publishers.

Hart, D. D., & Bell, K. P. (2013). Sustainability science: A call to collaborative action. *Agricultural and Resource Economics Review, 42*(1), 75–89.

Hartley, L., Anderson, C., Berkowitz, A., Moore, J., Schramm, J., & Simon, S. (2011). *Development of a grade 6–12 learning progression for biodiversity: An overview of the approach, framework, and key findings.* Paper presented at the National Association for Research in Science Teaching Annual Meeting, Orlando, FL.

Harvey, M., & Pilgrim, S. (2011). The new competition for land: Food, energy, and climate change. *Food Policy, 36*(Supplement 1), S40–S51. https://doi.org/10.1016/j.foodpol.2010.11.009

Holling, C. S. (1973). Resilience and stability of ecological systems. *Annual Review of Ecology and Systematics, 4*(1), 1–23.

Hufnagel, E., Kelly, G. J., & Henderson, J. A. (2017). How the environment is positioned in the next generation science standards: A critical discourse analysis. *Environmental Education Research, 1*, 1–23.

Jackson, J. K., & Fisher, S. G. (1986). Secondary production, emergence, and export of aquatic insects of a Sonoran Desert stream. *Ecology, 67*(3), 629–638.

Jianguo, W. (2013). Key concepts and research topics in landscape ecology revisited: 30 years after the Allerton Park workshop. *Landscape Ecology, 28*(1), 1–11.

Kincheloe, J. L. (2008). *Critical pedagogy primer.* New York, NY: Peter Lang.

Kraemer, A. L. (2012). *Encountering corporate responsibility: Mining, development and conservation in south eastern Madagascar.* SOAS, University of London.

Labaree, D. F. (2008). The winning ways of a losing strategy: Educationalizing social problems in the United States. *Educational Theory, 58*(4), 447–460.

Landscape Conservation Cooperative Network. Retrieved from https://lccnetwork.org/

Latour, B. (2012). *We have never been modern.* Cambridge, MA: Harvard University Press.

Laughlin, R. B. (2008). *A different universe: Reinventing physics from the bottom down.* New York, NY: Basic Books.

Levin, S. A. (2005). Self-organization and the emergence of complexity in ecological systems. *Bioscience, 55*(12).

Levinas, E. (1979). *Totality and infinity: An essay on exteriority.* Boston, MA: Nijhoff.

Lévinas, E., & Robbins, J. (2001). *Is it righteous to be?: Interviews with Emmanuel Lévinas.* Stanford, CA: Stanford University Press.

Lorenzoni, I., & Pidgeon, N. (2006). Public views on climate change: European and USA perspectives. *Climatic Change, 77*(1), 73–95.

Loza, A. (2016). There are only few places where organic bananas can be grown. *Fresh Plaza.* Retrieved from http://www.freshplaza.com/article/158803/There-are-only-few-places-where-organic-bananas-can-be-grown

Mancus, P. (2007). Nitrogen fertilizer dependency and its contradictions: A theoretical exploration of social-ecological metabolism. *Rural Sociology, 72*(2), 269–288.

Manel, S., Schwartz, M. K., Luikart, G., & Taberlet, P. (2003). Landscape genetics: Combining landscape ecology and population genetics. *Trends in Ecology & Evolution, 18*(4), 189–197. https://doi.org/10.1016/S0169-5347(03)00008-9

McFarlane, D. A., & Ogazon, A. G. (2011). The challenges of sustainability education. *Journal of Multidisciplinary Research, 3*(3), 81.

Metzger, E. P., & Curren, R. R. (2017). Sustainability: Why the language and ethics of sustainability matter in the geoscience classroom. *Journal of Geoscience Education, 65*(2), 93–100.

Miller, T. R. (2013). Constructing sustainability science: Emerging perspectives and research trajectories. *Sustainability Science, 8*(2), 279–293. https://doi.org/10.1007/s11625-012-0180-6

Mohan, L., Chen, J., & Anderson, C. W. (2009). Developing a multi-year learning progression for carbon cycling in socio-ecological systems. *Journal of Research in Science Teaching, 46*(6), 675–698.

Mueller, M. P., & Tippins, D. J. (2012). Citizen science, ecojustice, and science education: Rethinking an education from nowhere. In *Second international handbook of science education* (pp. 865–882). Dordrecht, The Netherlands: Springer.

Ogden, L., Heynen, N., Oslender, U., West, P., Kassam, K.-A., & Robbins, P. (2013). Global assemblages, resilience, and earth stewardship in the Anthropocene. *Frontiers in Ecology and the Environment, 11*(7), 341–347.

Olsson, L., Jerneck, A., Thoren, H., Persson, J., & O'Byrne, D. (2015). Why resilience is unappealing to social science: Theoretical and empirical investigations of the scientific use of resilience. *Science Advances, 1*(4), e1400217.

Oreskes, N., & Conway, E. M. (2011). *Merchants of doubt: How a handful of scientists obscured the truth on issues from tobacco smoke to global warming.* New York, NY: Bloomsbury Publishing USA.

Pedretti, E., & Nazir, J. (2011). Currents in STSE education: Mapping a complex field, 40 years on. *Science Education, 95*(4), 601–626. https://doi.org/10.1002/sce.20435

Polanyi, K. (1957). *The great transformation.* Boston, MA: Beacon Press.

Polis, G. A., Anderson, W. B., & Holt, R. D. (1997). Toward an integration of landscape and food web ecology: The dynamics of spatially subsidized food webs. *Annual Review of Ecology and Systematics, 28*(1), 289–316. https://doi.org/10.1146/annurev.ecolsys.28.1.289

Puko, T. (2017). Trump Administration takes steps to replace Obama Clean Power Plan. *The Wall Street Journal.* Retrieved from https://www.wsj.com/articles/trump-administration-takes-steps-to-replace-obama-clean-power-plan-1507158937

Resilience. Retrieved from https://www.resalliance.org/resilience

Rittel, H. W., & Webber, M. M. (1973). Dilemmas in a general theory of planning. *Policy Sciences, 4*(2), 155–169.

Roe, E. (2016). Environmental ethics. In *International encyclopedia of geography: People, the earth, environment and technology.* London, UK: Wiley.

Roth, W.-M., & Desautels, J. (2002). Science education as/for sociopolitical action: Charting the landscape. *Counterpoints, 210,* 1–16.

Ruppert, J., & Duncan, R. G. (2017). Defining and characterizing ecosystem services for education: A Delphi study. *Journal of Research in Science Teaching, 54*(6), 737–763. https://doi.org/10.1002/tea.21384

Seufert, V., Ramankutty, N., & Foley, J. A. (2012). Comparing the yields of organic and conventional agriculture. *Nature, 485,* 229. https://doi.org/10.1038/nature11069

Sharma, A. (2016). STEM-ification of education: The zombie reform strikes again. *Journal for Activist Science and Technology Education, 7*(1), 42–51.

Shear, M. D. (2017). Trump will withdraw US from Paris climate agreement. *The New York Times.* Retrieved from https://www.nytimes.com/2017/06/01/climate/trump-paris-climate-agreement.html

Shepardson, D. P., Roychoudhury, A., & Hirsch, A. S. (Eds.). (2017). *Teaching and learning about climate change: A framework for educators*. New York, NY: Taylor & Francis.

Sherkat, D. E., & Ellison, C. G. (2007). Structuring the religion-environment connection: Identifying religious influences on environmental concern and activism. *Journal for the Scientific Study of Religion, 46*(1), 71–85.

Smith, K., Rowe, D., & Vorva, M. (2015, December). The status of education for sustainable development (ESD) in the United States: A 2015 report to the US Department of State: International Society of Sustainability Professionals.

Spirn, A. W. (2014). Ecological urbanism: A framework for the design of resilient cities (2014). In F. O. Ndubisi (Ed.), *The ecological design and planning reader* (pp. 557–571). Washington, DC: Island Press/Center for Resource Economics.

Sustainability Science. Retrieved from http://sustainability.pnas.org/

Tilman, D., Socolow, R., Foley, J. A., Hill, J., Larson, E., Lynd, L., ... Somerville, C. (2009). Beneficial biofuels—The food, energy, and environment trilemma. *Science, 325*(5938), 270–271.

Turner, M., & Cardille, J. (2007). Spatial heterogeneity and ecosystem processes. In J. Wu & R. Hobbs (Eds.), *Key topics in landscape ecology, Cambridge studies in landscape ecology* (pp. 62–77). Cambridge, UK: Cambridge University Press. https://doi.org/10.1017/CBO9780511618581.005

Tyedmers, P., & Pelletier, N. (2007). Biophysical accounting in aquaculture: Insights from current practice and the need for methodological development. *Comparative assessment of the environment costs of aquaculture and other food production sectors: methods of meaningful comparisons*. FAO. Rome. FAO Fisheries Proceedings (10), 229–241.

UNESCO. (2005). *The DESD at a glance*. Retrieved from Paris: http://unesdoc.unesco.org/images/0014/001416/141629e.pdf

Vandermeer, J. H., & Perfecto, I. (2005). *Breakfast of biodiversity: The political ecology of rain forest destruction*. Oakland, CA/New York, NY: Food First Books.

Walker, B., Salt, D., & Reid, W. (2012). *Resilience Thinking: Sustaining Ecosystems and People in a Changing World*. Washington, DC: Island Press.

Warwick, S. I., Beckie, H. J., & Hall, L. M. (2009). Gene flow, invasiveness, and ecological impact of genetically modified crops. *Annals of the New York Academy of Sciences,1168*(1),72–99.https://doi.org/10.1111/j.1749-6632.2009.04576.x

Welldon, J. E. C. (1897). *The Nicomachean ethics of Aristotle*. London, UK: Macmillan and Company, limited.

Whatmore, S. (1997). Dissecting the autonomous self: Hybrid cartographies for a relational ethics. *Environment and Planning D: Society and Space, 15*(1), 37–53.

Whiteside, K. H. (2002). *Divided natures: French contributions to political ecology*. Cambridge, MA: MIT Press.

Index[1]

A
Abiotic factors, 135, 142
Action over actors, 66, 174
Actor-network theory (ANT), 33
Air pollution, 156, 162
Allspaw, K., 12
American Association for the Advancement of Science (AAAS), 37, 50
Ampatzidis, G., 155
Anderson, C. W., 9, 155, 183, 194
Anthromes, 71, 114n2
Anthromorphism, 189
Anthropocene, 5, 12, 35, 39, 73, 197
Anthropocene Epoch, 2, 12, 39, 176, 180, 183, 184, 186
Archer, K., 30
Aronowitz, S., 198

B
Bäckstrand, K., 78
Balance of nature concept, 10, 11, 24, 27, 28, 32, 34, 145, 155, 156, 178
Barad, K., 38
Barrett, G. W., 81n5
Bastian, M., 191
Bekoff, M., 28, 110
Benchmarks for Science Literacy, 51, 55
Bergesen, A. J., 105
Binder, C., 192
Biodiversity, 21, 55, 92, 96, 108, 123, 184, 185
Biomes, 12, 71, 88, 99, 100, 114n2, 142, 151, 153
Biotic factors, 135, 142
Black Death, 24
Bonnett, M., 154
Bots, P., 192
Bradshaw, G. A., 28, 110

---

[1] Note: Page numbers followed by 'n' refer to notes.

© The Author(s) 2018
A. Sharma, C. Buxton, *The Natural World and Science Education in the United States*, https://doi.org/10.1007/978-3-319-76186-2

Braun, B., 39
Bunker, S. G., 111
Business-as-usual emissions, 3
Buxton, 144, 160

C
Capitalist modes of production, 177
Carbon dioxide, 1, 3, 107, 139, 159
  removal, 3
Carbon TIME, 12
Carnegie Corporation of New York, 49, 50
Cartesian-Newtonian perspective, 175, 177
Castree, N., 39
Cells and Heredity, 135
"Characteristics of Science" standards, 55
Cheng, J. C., 10
Citizen science, 8, 199
Civilizational collapse, 2
Classroom discourse, 11, 22, 40, 65, 93, 129, 153, 154
Classroom exchanges, 129
Classroom-based research, 93
Clean Power Plan, 169
Climate change, 3, 11, 21, 66, 150, 180
Cobern, W. W., 6, 10, 151, 152
Communicative action, 128
Conservation-as-development paradigm, 177
Content, 187, 191, 192
Cost-benefit ratios, 75, 77, 78
"Covering Law" model, 135
Criterion-Referenced Competency Test (CRCT), 130
Critical discourse analysis (CDA), 47–48, 90–92, 113, 123, 170
Critical realist political ecology, 32
Cronon, W., 30
Curriculum material, 170

D
Dedeurwaerdere, T., 36, 187
Deepwater Horizon Oil Spill, 160
Deforestation, 158
Democratic decision-making, 157
Democratic ecologism, 189–191
Description-explanation-generalization, 128
Dewey, J., 25
Dialogic-authoritative communication, 128
Dirzo, R., 52
Dominant discourses, 4, 125, 176
Dominant Social Paradigm (DSP), 4
Dryzek, J. S., 74, 75
Dualisms, 12, 25–29, 31, 33, 34, 39, 40n1, 144, 153
Duncan, R. G., 9, 192
Dyson, A. H., 122

E
Earth systems, 72
  change, 61
The Earth's Ecosystems, 100
Ecocentrism, 189
Ecofeminism, 27
Ecojustice, 8
Ecological discourses, 163
Ecological flows, 194
Ecological modernization, 74–78, 175, 176
Ecological organization, 138
Ecological succession, 172
Ecologism, 189
Ecology, 5–8, 11–14, 28, 29, 31, 32, 36, 62, 71–73, 92, 93, 97, 113, 121, 123, 130, 132, 133, 141
Eco-managerialism, 76

Ecosystems
  degradation, 76
  ecology, 73, 175
  organisms in, 63, 135
  services, 9, 72
Education for Sustainable
  Development (ESD), 181–183
Educational reform, 46, 181
Enacted curriculum, 121–146,
  177–178
Energy flows, 195
Environmental action, 2, 4, 65, 78,
  79, 87, 97, 169, 173, 175
  individualization of, 108–109
Environmental change, 31, 75, 111
Environmental discourse, 73–74,
  174–177
Environmental education, 7, 8, 10,
  11, 73
  reforming, 181–183
Environmental governance, 197
Environmental issues, 8, 9, 62, 66, 67,
  73, 76, 77, 79, 87, 89, 107, 111,
  130, 144, 149, 150
  students' understanding of,
  156–163
Environmental Literacy Project,
  183–184
Environmental problems, 107
Environmental science, 130, 141
Environmental Science Literacy
  Project, 200
Environmental threat, 105–108
Environmental transformation, 31
Environmentalism, 2, 3
Ergazaki, M., 155
Ethical pluralism, 191
Ethical relationship, 26
Ethical stance, 6, 14, 36, 187, 189, 190
Ethical-political dimensions, 185
Ethico-moral decision-making, 8
Ethics, 189, 190

Ethnography, 123
Eutrophication, 100
Examination questions, 132
Exclusion backgrounding (EB), 96
Exclusion suppression (ES), 96
Existential assumptions, 48
Existential processes, 95
External nature, 21

F
Fairclough, N., 47, 59, 60, 90
Fang, Z., 89
Feinstein, N. W., 9, 170, 176
Filotas, E., 194
Fisher, S. G., 194
Flux of nature metaphor, 28
Flyvbjerg, B., 188
Folke, C., 198
Food chain, 134
Formsinho, J., 190
A Framework for K-12 Science Education, 51–53, 69
Freudenburg, W. R., 105
Fundamentals of Ecology, 81n5

G
Gardner, H., 171
Genishi, C., 122
Georgia Performance Standards
  (GPS), 13, 51, 55–58, 62, 69, 79,
  81n2, 127, 133, 171
Georgia science standards, 13, 74, 76
Georgia Science Teachers Association,
  51, 52
Georgia Standards for Excellence
  (GSE), 13, 51, 52, 55, 57–59,
  62, 63, 65, 69, 75, 173
*Georgia: Holt Science and Technology:
  Life Science*, 88, 92, 171, 172
Geoscience processes, 70

Gillson, L., 28
Gilmour, Rebecca, 121
  school context, 124–125
  school science discourse, 126–129
  seventh-grade science class, 125–126
Giroux, H. A., 198
Global climate change, 53
Global discourses, 92
Grant, D. S., 105
Green capitalism, 74, 79, 80, 175
Green governmentality, 78–80, 175, 176
Gruenewald, D. A., 7

H
Habits of Mind standards, 51, 55
Hahn, T., 198
Halliday, M., 91, 95, 97, 113
Hansen, J., 1
Haraway, D. J., 28
Harvey, D., 30
Hempel, L. C., 111
Henderson, J. A., 67, 172
Hinkel, J., 192
Hispanics, 124
Holocene Epoch, 35
Holt, R. D., 194
Hovardas, T., 170
HS-ESS3-3, 66, 68, 75
Hufnagel, E., 67, 170, 172, 173
Human action, 64, 68
Human agency, 97–105, 110
Human civilizations, 39
Human culpability, 100, 101
Human exceptionalism, 4, 26, 154, 156, 163, 179, 189
Human impact, 61, 172
Human population, 10, 63, 66, 74, 75, 79, 192
Human reason, 65, 66

Human-nature dualism, 39
Human-nature interactions, 97–105
Human-nature relationships, 154, 174

I
Ideational metafunction, 91
Information flows, 196
Initiation-response-evaluation (IRE), 128
Initiation-response-feedback-response-feedback (IRFRF), 128
Instructional strategies, 131
Integrated socioecological systems, 191
Intended Curriculum, 45–81, 87–113, 170, 171
  Georgia schools, 51
Interactive-non-interactive communication, 128
International Association for the Evaluation of Academic Achievement (IEA), 37
Interpersonal metafunction, 91
Intrinsic nature, 21
IPAT equation, 79

J
Jackson, J. K., 194
Jones, A. W., 105
Judeo-Christian religious discourses, 4

K
Kelly, G. J., 67, 88, 172
Kirchgasler, K. L., 9
Knowledge production, 33
Koch, A. M., 24
Korfiatis, K., 170
Kyoto Protocol, 181

L

Ladle, R. J., 28
Landscape Conservation Cooperatives (LCC), 198
Language-in-use discourse, 91
Latour, B., 25, 28, 38
Learning progressions, 184
Lee, V. R., 93, 144
Levin, S., 193
Levinas, E., 187
Levins, R., 27
Lewontin, R., 27
Lin, H., 10
Linguistic smokescreen, 105
Little Creek Middle School, 121, 123–125, 149
Liu, S., 10
Logical implications, 48
Lorenzoni, I., 162
Loughland, T., 151
Lövbrand, E., 78

M

Mainstream media, 156
Manufacture consent, 68
Marginalized communities, 173
Market-based exchanges, 112
Martin, J., 113
Material clauses, 96, 101, 102
Material flows, 195
Material processes, 95, 101
Materio-spatial world systems analysis, 32
Mazid, B. M., 47
McKibben, Bill, 1
Mental processes, 95
Mesolevel qualitative documents analysis, 97
Metabolic relationship, 143
Michigan State University (MSU), 10, 12, 183, 191, 200
Microlevel transitivity analysis, 97
Middle school science classrooms, 109
Middle-grade science classroom, 93
Middle-grade science standards, 112
Middle-grade science textbook, 109
Modern age, 23–27
Modern Constitution, 25, 26, 28
Monroe, M. C., 10
Moral relationship, 26
Mortimer, E. F., 127, 128
MS-ESS3-1, 70, 74, 81n3
Mutual sustainability, 151, 179
Mutualism, 135

N

National Research Council (NRC), 48, 50
National Science Education Standards, 49
National Science Teachers Association (NSTA), 50
Natural resources, 76
Natural sciences, 25
Natural systems, 22
Naturalistic generalization, 122
Natural-social interactions, 103
Natural-social relationships, 97, 100, 110
Nature
  abstraction, 131–135
  amodern view of, 30–34
  conceptions of, 23–27
  Human-Free World, 140–143
  in Science Textbook, 87–113
  mass-mediated representations of, 150
  modern science and, 27–30
  in school science, 170, 171, 177–181
  in science teaching, 121–146
  Stable World, 135–139
  Students' Understanding of, 150–156

Nature of Science, 51
Nature Qua abstraction, 131
Naturecultures, 186
Nature-human relationships, 74
Nature-social dualism, 27, 29, 172, 175
Neoliberalism, 112
New London Group, 95
Next Generation Science Standards (NGSS), 9, 13, 46, 50–55, 57, 58, 61–64, 67, 69, 70, 72, 74–76, 78, 79, 81, 81n3, 133, 171–173, 176
Ngram Viewer, 21
Nielsen, J. A., 152
No Child Left Behind Act of 2001, 49
Noncentered democratic ecologism, 190
Nonpoint-source water pollution, 105
Nonprofit education reform organization, 40n2
Norberg, J., 198

O
Odum, E. P., 81n5
Olsson, P., 198
Ontology, 61, 69, 171–173
Opinion polls, 107
Organic bananas, 185
Organisation for Economic Co-operation and Development (OECD), 37
Organism flows, 195
Othering, 159, 180
Otherness, 154, 163, 179
Ozone depletion, 150

P
Pahl-Wostl, C., 192
Paris Agreement, 181
Paris Climate Accord, 1, 169

Payne, P., 153
Pellow, D. N., 103
Per-capita consumption, 79
Perfecto, I., 185
Petocz, P., 151
Pidgeon, N., 162
Place-based science education, 8
Pointon, P., 154
Polis, G. A., 194
Politics of men, 26
Portland Baseline Essay Project, 77, 81
Portland School District curriculum, 78
Potter, E. F., 93
Power of the mind, 24
Prairie ecosystem, 137
Praxis, 187, 199
Process over cause, 66, 174
Propositional assumptions, 48

Q
Quigley, C., 12

R
Rainforest Trust, 184
Received Curriculum, 149–164, 178–181
Recovery process, 155
Reid, A., 151
Relational processes, 95
Resilience-based research, 197
Review game, 136
Rittel, H., 186
Rockstrom, J., 82n6
Roe, E., 190
Rosser, S. V., 93
Roth, W.-M., 188
Rowe, D., 182
Ruppert, J., 9, 170, 192

S

Schnaiberg, A., 103, 111
School curriculum, 73
School education, 39
School science, 7–12
  content, 36
  discourse, 37–39, 89, 110, 112, 123, 125, 128–130, 136, 139, 144, 170
  ethics, 36
  praxis, 36
  rethinking, 34–40
  sustainability science, 35–37
  curricula, 189
School-based education, 188
Science classrooms, 10, 22
  settings, 122
Science content standards, 47–52
Science curriculum, 139, 190
Science discourse, 145
Science education, 7, 111, 169–200
  age of wicked problems, 184–199
Science education community, 6
Science educators, 169, 176
Science knowledge, 64, 68
Science of things, 26
Science standards, 170
  Georgia-owned and Georgia grown, 51
  world and, 53–81
Science teachers, 178
Science teaching apprenticeship, 126–127
Science textbooks, 89
Science-Technology-Society (STS) education, 8, 36
Scientific and environmental discourses, 178
Scientific discourse, 69–73, 145, 174–176
Scientific ideas, 129
Scientific illiteracy, 5
Scientific inquiry, 65
Scientific knowledge, 151
Scientific literacy, 49
Scientific Method, 129
Scott, P. H., 127, 128
Self-organization, 193
Self-organizing complex system, 35
Shepardson, D. P., 153
Smith, K., 182
Smith, N., 30
Social ecology, 27
Social hierarchy, 60
Social metabolism, 25
Social technology tools, 59
Social world, 70
Social-ecological justice, 6
Social-ecological resilience, 196
Social-ecological systems, 31, 192
Social-nature dualism, 28, 29, 31, 34, 39
Social-nature dualist ontology, 10
Societal discourses, 73, 112, 122, 164
Societal issues, 3, 73
Society–nature dualism, 33
Socioecological context, 77
Socioecological systems, 192, 193
Socioscientific Issues (SSI), 8
Socioscientific Studies Issues-based (SSI) Education, 36
Stability, 61
Standards-based reforms, 46
State education agencies, 49
State of Georgia, 92, 107, 125
State Textbook Advisory Committee, 92
State-of-the-art understanding, 170
Status quo, 181
STEM education reform, 200
STEM Funders Network, 37
Stone-Jovicich, S., 32

Student learning, 88
Students gender distribution, 126
Substantivist ontology, 37
Sustainability science, 33–37, 39
Sustainability science-based framework, 169–200
Sustainability science-oriented science education, 200
Systemic functional linguistics (SFL), 89–91, 94
Systems theory, 62

T
Teacher interventions, 129
Technocentric value system, 173
Technocentrism, 67, 174
Technocratic-economic logic, 77
Technoscience, 174
Textbook adoption, 112
Textual metafunction, 91
Textually oriented approach to critical discourse analysis (TODA), 47, 90, 91
Thinking curriculum, 48
Tragedy of commons, 4, 109
Transitivity, 94
Tsurusaki, B. K., 10, 155
Turin, D. R., 2

U
Unemployment rate, 124

United States, 7, 10, 48, 73, 123, 144, 153, 155, 157, 160, 169, 178, 180
Universal nature, 21
Unrestrained instrumental rationality, 26
US Department of Education, 37

V
Value assumptions, 48
Values, 64–68, 173–174
Vandermeer, J. H., 185
Vocabulary term, 131
Vorva, M., 182

W
Walker, B., 193
Warm-up questions, 131
Water wars, 158
Webber, M., 186
Weinberg, A., 103
Western civilization, 23
Wixson, K. K., 46
Wodak, R., 90
World, perspectives on, 174
Writing process, 81n3

Y
Yellowstone National Park, 67

The manufacturer's authorised representative in the EU is Springer Nature Customer Service Centre GmbH, Europaplatz 3, 69115 Heidelberg, Germany. If you have any concerns regarding our products, please contact ProductSafety@springernature.com

Printed and bound by CPI Group (UK) Ltd, Croydon, CR0 4YY
23/03/2026
02076739-0002